£34.65

D0301740

R. J. Whittaker
School of Geography
University of Oxford
15. 5. 90.

MOUNT ST. HELENS 1980

Published in association with the Pacific Division
American Association for the Advancement of Science

Mount St. Helens 1980

Botanical Consequences of the Explosive Eruptions

Edited by
David E. Bilderback

UNIVERSITY OF CALIFORNIA PRESS
Berkeley   Los Angeles   London

University of California Press
Berkeley and Los Angeles, California

University of California Press, Ltd.
London, England

Library of Congress Cataloging in Publication Data
Main entry under title:

Mount St. Helens, 1980.

   Papers presented at a symposium held in June, 1981, during the 62nd Annual
Meeting of the Pacific Division of the American Association for the Advancement
of Science and sponsored by the Pacific Section of the Botanical Society of
America and the Western Section of the Ecological Society of America.
   Includes index.
   1. Botany—Washington (State)—Saint Helens, Mount, Region—Ecology—
Congresses. 2. Saint Helens, Mount—Eruption, 1980—Congresses. 3. Plant
communities—Washington (State)—Saint Helens, Mount, Region—Congresses.
4. Vegetation dynamics—Washington (State)—Saint Helens, Mount, Region—
Congresses. I. Bilderback, David E. II. Botanical Society of America. Pacific
Section. III. Ecological Society of America. Western Section. IV. Title: Mount
Saint Helens, 1980.
QK192.M68   1986 581.5'222 85-24650
ISBN 0-520-05608-6

Printed in the United States of America

1 2 3 4 5 6 7 8 9

CONTENTS

Usful comparison (handwritten annotation)

PREFACE

On May 18, 1980, the eruption of Mount St. Helens began with a massive avalanche of rock and snow and a lateral blast of rock and superheated steam that leveled approximately 300 km² of forested lands. Subsequent deposition of pyroclastics, mudflows, and tephra further transformed the landscape around the volcano. To the east, the drifting volcanic plume deposited varying amounts of ash on the forest and steppe communities of eastern Washington, northern Idaho, and western Montana.

The eruption of Mount St. Helens was unique in that it occurred near major academic, governmental, and industrial centers of ecological research. The area around the mountain became accessible to many scientists interested in the immediate and long-term effects of volcanism on vegetation and its subsequent recovery. Farther from the volcano, scientists began investigations on the effects of tephra deposition on ecosystems. Never in human history have the effects of a volcanic eruption been so intensively investigated by so many highly qualified scientists.

At the 62d Annual Meeting of the Pacific Division of the American Association for the Advancement of Science, held on June 14–19, 1981, the Pacific Section of the Botanical Society of America and the Western Section of the Ecological Society of America co-sponsored a major symposium on the biological effects of the Mount St. Helens eruption. At that symposium, scientists presented their initial findings concerning the effects of the eruption on plant communities. Those reports, in turn, have been compiled in this volume. It is hoped that this volume will provide important baseline information and will stimulate further investigations on the long-term effects of volcanism on vegetation and ecosystems.

The editor wishes to thank Dr. David Wagner, Department of Biology, University of Oregon, for organizing a most successful symposium; Ms. Helena Chambers for her able editorial assistance; and Dr. Alan Leviton, Executive Director, Pacific Division of the American Association for the Advancement of Science, for his continued encouragement and support.

DAVID E. BILDERBACK

1

Ecological Effects of the Eruption of Mount St. Helens: An Overview

Frederick J. Swanson

The diverse effects of volcanic activity on ecosystems described in this volume reflect the complexity of the volcanic events associated with the May 18, 1980, eruption of Mount St. Helens. During the eruption, a huge avalanche of rock and snow, a directed lateral blast of rock and superheated steam and gases, repeated pyroclastic flows, ashfall, and mudflows dramatically transformed the mountain and the surrounding landscape. Farther from the mountain, airborne tephra was deposited on the various ecosystems of eastern Washington, northern Idaho, and western Montana. Each volcanic and hydrologic event would be expected to have distinctive impacts on ecosystems and influence subsequent ecosystem recovery in various ways. For this reason, analysis of the ecological responses to the eruption must be based on a detailed understanding of the volcanic and hydrologic processes that initially perturbed the ecosystems. Furthermore, recovering ecosystems undergo rapid physical and biological changes, and interactions among plants, animals, and geomorphic processes are particularly dynamic. Consequently, the response of any one component of a system can be best understood in terms of the overall behavior of the ecosystem and the landscape.

The debris avalanche area presents especially complex problems in sampling recovering vegetation. The surface materials of these deposits exhibit great heterogeneity in terms of the topography, the lithology, and the abundance of residual organic matter. The surface has subsequently been altered by fluvial erosion and deposition and by mudflows. Revegetation has begun at the margins of the deposit where transported soil blocks containing propagules of plants, mycorrhizal fungi, and nutrients are concentrated. Plant recovery has been slow where fluvial processes have repeatedly altered ecosystems on the avalanche surface. Consequently, plant communities are forming a complex mosaic determined mainly by the landforms and geomorphic processes occurring at each microsite.

In the blast zone, complex interactions occur among hillslope erosion processes, living plants, and dead woody debris. Timber felled by the lateral blast retards sheet and rill erosion of tephra deposits by slowing the flow of water and sediment downslope. Initial recovery, however, has been dominated in many areas by sprouting of below-ground parts of residual plants where erosion of the tephra deposits has exposed the old soil surface. Elsewhere in the blast zone, ecosystem reestablishment has been sparse, consisting of a few colonizing organisms or a few residual plants and animals emerging from below ground or from snowbanks present on the day of the eruption. These small oases of organisms, however, are facilitating establishment of small animals and other plant species from seed or spores by providing shade, food, seed-trapping sites, and litter.

Farther from the volcano, airborne tephra deposited on forest and steppe communities has had complex effects on ecosystem interactions among plants, animals, and the altered physical environment. Of particular interest is the long-term effect that the ash layer will have on seed germination dynamics and the buried, delicate cryptogamic crusts of the arid lands of eastern Washington.

A comprehensive investigation of the ecological recovery occurring at Mount St. Helens will provide a unique opportunity to gain a greater understanding of the evolution of ecosystems around other stratovolcanoes of the Cascade mountain range of the Pacific Northwest and elsewhere in the world. In turn, observations of the stages and patterns of recovery of volcanically disturbed ecosystems elsewhere in the Cascade range may be of important value in predicting the course and rate of future ecosystem recovery around Mount St. Helens.

Plant Life on Mount St. Helens before 1980

Arthur R. Kruckeberg

ABSTRACT

Approximately 315 species of vascular plants were known from Mount St. Helens, Washington, before the eruption of 1980: 17 fern taxa, 13 conifers, 68 monocots, and 217 dicots. The flora was depauperate in comparison to other volcanoes of the Cascade mountain range. No taxa were endemic to Mount St. Helens nor were there any endangered or rare taxa. Many of the plant species found on other Pacific Northwest volcanoes were absent; however, the xeric pumice slopes did support species also found growing on the pumice of nearby volcanoes. The parkland and timberline vegetation types were underdeveloped on Mount St. Helens and had an atypical mix of low and high elevation conifers. The xeric, unstable pumice and the recency (early 1800s) of previous volcanic activity may have contributed to the atypical flora and vegetation of this mountain.

INTRODUCTION

The proximity of Mount St. Helens to major population centers and its spectacular eruptive displays in 1980 should not make us forget that this most active local volcano is only one of many in the Pacific Northwest and a latecomer among the volcanoes of the Pacific "ring of fire." Any questions regarding the uniqueness of the flora and vegetation and the possible reappearance of these unique attributes should be asked within the broader context of the plant life on volcanoes in general. The next three sections cover a general discussion of plant life on volcanoes, followed by a more specific discussion of the plant life on Mount St. Helens before 1980.

Because the author's interests lie in floras of unusual substrates (Kruckeberg 1969a, 1969b), the floristic aspects of higher plants will be emphasized, with only scant attention paid to vegetation communities. The floras of lower plant groups that may be found on volcanoes are not discussed. The author has collected plants on many volcanoes of the Pacific Northwest, from Mount Lassen to Mount Baker, since 1950, and he made three collecting trips to Mount St. Helens before 1980 (Kruckeberg 1979).

ATTRIBUTES OF VOLCANOES AND THEIR PLANT LIFE

It seems likely that unique floras would evolve with structural and functional adaptations for life on volcanoes. Furthermore, those floras might display themselves as unique vegetation types or communities. To test these premises, we should first inventory the physical attributes of volcanoes that might promote special kinds of plant life. Climate, geographic location, substrate (rock and soil), altitude, and topography all vary among the many kinds of volcanoes and their global locations. The following paragraphs describe some of the variables that can affect the kinds of plant life found on volcanoes.

1. *Kinds and periodicity of volcanism.*   The explosive pyroclastic type of volcanism can yield a very different landscape from that of volcanoes with viscous lava flows (the so-called shield volcanoes). More-over, the time interval between periods of activity can vary tremendously. Mount St. Helens has erupted approximately once every 100 yr since A.D. 1300, whereas Glacier Peak has been quiescent since the end of the Pleistocene Epoch (~14,000 yr ago).

2. *Geographic location of the volcano.*   Tropical versus temperate and oceanic or coastal versus continental locations are spatial variables that will profoundly influence the climatic setting of the volcano. In addition, the proximity of a particular volcano to similar peaks or to floras in dissimilar habitats will affect the consequent restocking of a raw volcanic substrate with plants.

3. *Soils of volcanic origin.*   The physical and chemical properties of volcanic substrates can vary significantly (Ugolini and Zazoski 1979) despite the attributes that volcanic parent materials and soils have in common. Volcanic soils are derived from tephra (volcanic ash) and pumice or from lava and tuff. The two classes of parent material can produce similar soils, but each type can vary depending on the kind of volcano and its regional climatic setting. Soils developed on tephra generally have a low bulk density, a friable consistency, a high weight percentage of organic matter, a clay fraction dominated by amorphous

material, a high pH-dependent cation exchange capacity, an ability to fix phosphorus, and a high water content (Ugolini and Zazoski 1979). Deficiency of macro- and micronutrients is often encountered. Both the physical properties and chemical constituents of volcanic soils are critical to the combination of factors determining the quality and quantity of plant life on volcanoes.

4. *Biological factors associated with plant life on volcanoes.* Given the regional setting for a particular volcano, its proximity to other similar habitats (volcanic or otherwise) must be considered. Flora and fauna of adjacent regions may function as suitable sources of biological colonists. Volcanoes such as Mount St. Helens, with histories of frequent eruptive events and only partial destruction of their biotas, can provide their own recruits. Several workers have stressed the importance of "hold-over" plants in contrast to new colonizers (see Smathers and Mueller-Dombois 1974). Barring complete destruction of the volcano, both kinds of habitat reclamation, hold-over plants and new colonists, will undoubtedly occur.

There is no clear agreement on the sequence of events in colonization by plants from the beginning of revegetation to a steady-state system. There is little consensus regarding the question of which species are the first colonizers. Bryophytes have been high-ranking candidates, only to be rejected in favor of blue-green algae or bacteria. In some instances, vascular plants are clearly the initial colonizers (often as holdovers) of new volcanic terrain (Smathers and Mueller-Dombois 1974).

VOLCANO FLORAS AND VEGETATION: A WORLDWIDE OVERVIEW

Volcanoes are expected to have unique floras and vegetation types, because the transient nature of the habitat (if it includes recurrent volcanism) promotes unique adaptive responses. Physical and chemical conditions of soil and parent material create many challenging habitat conditions: exposure, drainage, soil albedo, chemistry of volcanic gases, microclimate, microtopography, and instability of slope. Expectations for plant adaptations to such harsh conditions seem justified. This brief review of the plant life on temperate and tropical volcanoes emphasizes only flora that have accumulated long after an eruptive event. Other papers in this volume deal with the more immediate posteruptive recovery of plant life.

TEMPERATE VOLCANO FLORAS

A substantial portion of the rich and endemic flora of Japan is intimately associated with volcanism (Numata 1974). Kawano (1971) emphasized the importance of volcanoes in creating alpine habitats such as the gravel barren on the island of Hokkaido; he lists 26 species endemic to the volcanic gravels, all but two (species of *Salix*) being herbaceous.

San Francisco Mountain of northern Arizona is an eroded volcano formed in the early Pleistocene. Although it is the highest peak in Arizona at 3844 m (12,600 ft), the mountain has lost 914 m (3,000 ft) altitude by erosion. It appears as a mesic island in the desert and is covered with montane and alpine plants with northern and Rocky Mountain affinities. Only two endemic species are recognized from the alpine zone of the mountain: *Senecio franciscanus* Greene and an undescribed *Pedicularis* (Little 1941). Several other alpine species have been proposed as new endemics but have been reduced to synonomy or varietal status. Little (1941) remarks that "a large number of endemic species is not to be expected on a geologically young volcano." This may be true for the arid southwest and western North America but seems not to be true for other parts of the world where speciation has occurred rapidly following volcanic activity.

TROPICAL VOLCANO FLORAS

Volcanoes are common in the tropics. Many reach up out of the wet tropic lowlands to spectacular heights, and their montane and alpine slopes display rich floras. The volcanoes of the Philippines and Malaysia are tropical extensions of the Pacific ring of fire and typify the tropical floristic diversity on volcanoes.

The Hawaiian islands were born of volcanism, and the great shield volcano of Mauna Loa is still active. The incidence of flowering plant endemism (94%) on the islands is higher than for any other area in the world (Carlquist 1970). Although all of this endemism originated because of the volcanic creation of the island chain, a lesser proportion of endemic species are the direct response to volcanic activity. The spectacular silverswords and their relatives, the woody tarweeds in the genera *Argyroxiphium*, *Dubautia*, and *Wilkesia*, are good examples of plants uniquely adapted to volcanoes. Siegel and Siegel (chapter 14, this volume) have indicated that silverswords may possess unique tolerances to volcanic mercury in addition to their other remarkable features.

An island archipelago of volcanoes also exists in East Africa. The great volcanoes of Mount Kilimanjaro, Mount Kenya, and the Ruwenzori Mountains support a varied flora rich in endemics, often with bizarre life forms (e.g., the arborescent herbs of *Senecio* and *Lobelia* [Lind and Morrison 1974]).

A final example from the tropics, the Galápagos islands, is a xeric archipelago on the equator formed entirely by volcanism. Not more than 1−2 million yr old, the islands are still intermittently eruptive. Although none of the volcanic peaks is of great height, there is a clearly defined zonation in the vegetation. Wiggins and Porter (1971) estimate that 32.5% of the Galápagos flora is endemic. Of the total 642 species, 228 are endemics distributed among the following taxonomic groups: 10 ferns, 43 apetalae, 88 gamopetalae, 70 polypetalae, and 17 monocots. In general, isolated oceanic islands created by volcanism are notoriously rich in unique biota.

PACIFIC NORTHWEST VOLCANOES
AND THEIR PLANT LIFE

All of the loftiest peaks of the Cascade range are volcanoes, mostly stratovolcanoes (Harris 1980) whose isolated summits usually exceed the average elevation of the nearby mountains. The sequence of volcanic peaks, numbering 14 cones and calderas, dominates the Cascade landscape from Mount Lassen and Mount Shasta in northern California to Mount Garibaldi in southwestern British Columbia. These great peaks are latecomers in the evolution of the Pacific Northwest landscape, all having achieved their present form during the Pleistocene and Holocene epochs. Their youthful ages, their discontinuous distribution from south to north through 20° latitude, and their surpassing heights make them unique habitats for particular biota. Here is truly an archipelago of high "islands" extending from northern California to British Columbia. There are phenomena of great biogeographic interest in this vast stage of habitat diversity.

The effects of volcanism, as well as the distribution and physiognomy of Pacific Northwest volcanoes, can be brought into focus with three groups of biogeographical questions.

1. If some plants are endemic to volcanoes, are they restricted to only one volcano or to several? Can taxa be endemic to several volcanoes?

There are many examples of endemic species on volcanoes of the Pacific Northwest. Some are restricted to single volcanoes; others have a wider distribution. Outstanding examples include the following (see Tables 2.1–2.3 for authors of taxa): *Botrychium pumicola* (Mount Mazama, Oregon; Mount Shasta, California), *Arenaria pumicola* (Mount McLoughlin to Mount Jefferson, Oregon), *Draba aureola* (Mount Lassen, California; Three Sisters and Diamond Peak, Oregon; Mount Rainier, Washington), *Smelowskia ovalis* (var. *congesta*, Mount Lassen; var. *ovalis* on several volcanoes of Oregon and Washington), *Tauschia stricklandii* (Mount Rainier, Washington; Columbia river gorge area), *Polemonium elegans* (Washington to southern British Columbia), *Pedicularis rainierensis* (Mount Rainier, Washington), *Aster gormanii* (Mount Jefferson to Breitenbush Lake, Oregon), *Hulsea nana* (Mount Lassen, California to Mount Rainier, Washington), and *Arnica viscosa* (Mount Shasta, California; Three Sisters, Mount Thielsen, and Mount Mazama, Oregon).

Another group of plants are often highly characteristic of volcanic habitats. Often called local or regional indicators, taxa of this group appear in volcanic areas with high fidelity and are often abundant there, but are not wholly restricted to volcanics. Prime examples include *Eriogonum pyrolaefolium* var. *coryphaeum*, *Polygonum newberryi*, *Silene suksdorfii*, *Anemone occidentalis*, *Saxifraga tolmiei*, *Lupinus lepidus* var. *lobbii*, *Collomia debilis* var. *larsenii*, *Aster alpigenus*, and *Luina stricta*. Note that the above two groups of plants are largely confined to the upper montane zone from parkland to alpine. Altitudes higher than the prevailing Cascade mode and volcanic substrates seem to foster occurrences of these plants. Table 2.1 shows endemic and indicator plants of Pacific Northwest volcanoes.

2. Are the nonendemic elements of volcanic floras modified in any infraspecific ways? That is, have they become ecotypically adapted to the volcanic environment?

Much of the regional flora of the Pacific Northwest has come to occupy volcanic habitats in ways in which the same species have occupied glacial till, glaciated valleys and slopes, older lavas, and other geomorphological features of the region. In other words, much of the flora seems broadly tolerant of the diverse substrates, as long as the climate (regional to meso- and microscale) is appropriate. This is particularly true for habitats of low to moderate elevations, such as the forested zones of the Cascades, the nonforested zones east of the Cascades, and the lowland cismontane regions. It is only at higher elevations (upper montane to alpine zones) that unique taxa or assemblages of plants seem to characterize our volcanoes.

Although there are few, if any, detailed observations to prove it, it seems likely that the taxa common to many habitats have become ecotypically modified on volcanoes. It would be expected that some subtle morphological and physiological attributes have evolved at the infraspecific level for survival on volcanoes. Thus it is highly probable that there are "volcano provenance" types of forest tree species. Foresters dealing with restocking of harvested timber can confirm this point. Whether or not such provenance types or ecological races in noncommercial species (woody and herbaceous) are adapted to volcano conditions per se or simply to the climatic regimes created by volcanoes is unknown.

Some distinctive plant communities have been recognized on volcanic substrates in the Pacific Northwest (Franklin and Dyrness 1973). The most frequent are the pumice communities that occur from Mount Lassen north to Mount Rainier. The most characteristic plants of these species-poor (simple) communities are *Eriogonum pyrolaefolium* var. *coryphaeum*, *Spraguea umbellata*, and *Lupinus lepidus* var. *lobbii*. The loose, porous pumice is found on flat terrain and on highly unstable slopes.

The *Festuca viridula* meadow type (Franklin and Dyrness 1973) characterizes much of the gently sloping terrain in the subalpine zone on the east side of Mount Rainier (e.g., Yakima Park). This same community type has *Lupinus latifolius* as a common co-dominant and can be seen elsewhere in volcanic areas of the Cascades.

Two community types result from recent volcanic activity (Franklin and Dyrness 1973). The highly irregular topography of Holocene lava flows supports a distinctive mixture of conifers, hardwoods, and herbaceous species. This edaphic nonconformity normally coincides with the *Abies amabilis* zone; instead, Douglas fir (*Pseudotsuga menziesii*) or subalpine fir (*Abies lasiocarpa*) are the usual tree dominants. Good examples can be seen in Oregon (the Nash Creek lava flows) and in southern Washington (the Big Lava Beds southwest of Mount Adams).

Mudflows (lahars) generated by volcanic events are widespread along the volcanic Cascades and often support successional stands of hardwoods or conifers. The mesic lahar at Kautz Creek on Mount Rainier, Washington, supports an alder–cottonwood–willow community that after 34 yr is presently being invaded by conifers. Franklin and Dyrness (1973) describe xeric communities on lahars in southern

Washington and Oregon that are typified by *Pinus contorta* and *Arcto-staphylos* (*A. nevadensis* or *A. uva-ursi*).

3. Does plant species diversity express itself uniquely on volcanoes? Does species diversity on volcanoes diminish with increasing latitude? Do the variables of size of surface area, volcano height, and recency and recurrence of volcanic activity affect the floristic composition of volcanoes?

No study has yet addressed these questions in detail; however, some plausible answers can be suggested. It is apparent that size and habitat diversity of volcanoes can be positively correlated with species diversity. Also, there appears to be a decrease in species diversity with latitude. For example, the flora of Mount Lassen is richer than that of Mount Baker (Cooke 1962). The element of time is also significant. The more recent or recurrent the episodes of volcanic activity, the more constrained is the display of floristic diversity. This is clearly evident for Mount St. Helens (as discussed in the next section).

These are only a few of the questions that could be asked by biologists interested in the volcanoes of the Pacific Northwest. Vegetative and reproductive traits that could be adaptive for life on volcanoes, biotic discontinuity between volcanoes, and the genesis and spread of plants on volcanoes are all problems that will stimulate biogeographers for years to come. The eruption of Mount St. Helens has certainly provoked inquiry in a dramatic manner.

THE FLORA AND VEGETATION OF MOUNT ST. HELENS PRIOR TO 1980

INTRODUCTION

Recounting the preeruptive state of Mount St. Helens on plant life might be considered a waste of time. The plant cover seemed at first to be almost completely eradicated, so it could be asked why a retrospective look is necessary? However, the vegetation of Mount St. Helens is far from eliminated. The preeruption vegetation has begun to return during the years of recovery. Indeed, many of the papers to follow in this symposium volume are testaments to the remarkable resilience of the volcano's flora. In this case, examining the past may aid us in predicting the future.

Unlike many loftier volcanoes of the Cascade range whose floras have been recorded as checklists or local floristic guides, the flora of

Mount St. Helens is almost without a historical record. If it were not for Harold St. John's checklist, published in 1976, we would have only a fragmentary record. St. John describes the activity of the first collector on the mountain: Frederick C. Colville, principal botanist for the U.S. Department of Agriculture. Colville visited the Pacific Northwest in 1898 and made collections on the southwest perimeter of the mountain on July 18–20, from Lake Merrill to the slopes of the mountain. It is not known if he attempted to scale the summit from his base camp at Three Buttes Camp (altitude 1403 m [4600 ft]). St. John, however, was able to locate several of the Colville specimens at the Smithsonian Institution.

This rediscovery of Colville's pioneering botanical trek to the mountain came many years after St. John himself had collected there. In fact, 51 years elapsed between St. John's 1925 reconnaissance of Mount St. Helens and his publication of "Flora of Mt. St. Helens, Washington" in 1976. His party of four, including George Neville Jones and Carl S. English, Jr., spent 12 days (August 2–13, 1925) collecting, first on the north side just above Spirit Lake (August 2–10) and then on the south side of the mountain (a shorter trip with Carl English only). One day of their stay on the north side was devoted to a climb to the summit. The St. John expedition inventoried the flora of the upper forested zone, including the western hemlock/western red cedar (*Tsuga heterophylla/Thuja plicata*) climax at Spirit Lake, the Pacific silver fir (*Abies amabilis*) zone, the treeline zone on the north slope, and the alpine of the mountain. St. John lists approximately 315 species: 17 ferns or fern allies, 13 conifers, 68 monocots, and 217 dicots.

Since the St. John expedition of 1925, no comprehensive flora of the mountain has appeared. Two Research Natural Areas have been established: Goat Marsh, a midelevation site with wetlands and adjacent forest, and Cedar Flats, a low elevation natural area exemplifying western red cedar stands on valley bottomlands. This has resulted in the publication of vegetation descriptions (Franklin and Wiberg 1979; Franklin et al. 1972) and unpublished plant species lists for the sites. Permanent plots set out by government and private foresters recorded dominant vegetation in the forested perimeter of the mountain (e.g., Hemstrom and Emmingham, chapter 7, this volume).

The last known botanical survey of Mount St. Helens was made in 1979. A party of 18 members of the Washington Native Plant Society spent 3 days on the north side of the mountain in the vicinity of Timberline Camp just south of and above Spirit Lake. On July 18, the party surveyed the treeline and lower alpine areas along the standard climbing route between Timberline Camp and the Sugar Bowl. The next day was spent in traversing the Plains of Abraham, from Windy Pass to the head of Ape Canyon. Significant floristic and vegetation features observed on this trip were published in the Society's newsletter *Douglasia* (Kruckeberg 1979). The full checklist of plants is available from the Society.

From the St. John study and the Washington Native Plant Society survey, an account of the vegetation and its floristic highlights can be given. This account will be confined mainly to descriptions of the north slope area from the upper montane to lower alpine zones.

VEGETATION DESCRIPTION

The upper montane and timberline areas of Mount St. Helens were not typical of nearby volcanoes such as Mount Hood, Mount Adams, and Mount Rainier. The remarkably distinctive Cascadian feature of subalpine parklands (the parkland subzone of the mountain hemlock [*Tsuga mertensiana*] zone of Franklin and Dyrness 1973), a spectacular terrain of tree clumps and intervening mountain meadow, was poorly developed at the Timberline Camp area. The transition from continuous forest to alpine was irregular and incomplete. Moreover, tree species composition was atypical for this zone. Rather than a grouping of the typical subalpine fir (*Abies lasiocarpa*), mountain hemlock (*T. mertensiana*), Alaska cedar (*Chamaecyparis nootkatensis*), or whitebark pine (*Pinus albicaulis*), timberline was an odd mixture of low elevation and upper montane conifers. In order of abundance were, first, lodgepole pine (*P. contorta*), subalpine fir, mountain hemlock, noble fir (*Abies procera*), and, finally, Pacific silver fir. Douglas fir, western hemlock, and western white pine (*P. monticola*) were least abundant. In this odd assemblage of conifers were *Populus trichocarpa* and *Alnus sinuata*—the only common hardwoods. Although alder is not unexpected at timberline, black cottonwood (a plant of low elevations) was out of place. Stunted individuals of this hardwood could be found as part of the atypical "Krummholz," along with Douglas fir, lodgepole pine, or subalpine fir. One must remember that this north slope was underlain by recent pumice fallout from earlier eruptions in the nineteenth century (1802, according to Lawrence 1954); thus it may not be surprising that a community structure more typical of the Cascadian upper montane had not yet taken form. Moreover, the pumice substrate—showing little soil-forming tendencies—appeared to form a xeric habitat. In addition to the exceptional composition of the tree cover, the shrub and herb layer was also unusual. *Lupinus latifolius* formed dense homogeneous swards on the open forest floor. In treeless areas on ridges or openings in the stunted woods, a pumice flora was dominated by *Lupinus lepidus* var. *lobbii*, *Penstemon cardwellii*, *Eriogonum pyrolaefolium* var. *coryphaeum*, *Agrostis diegoensis*, and *Luetkea pectinata*. *Spiranthes romanzoffiana* was locally frequent. *Phyllodoce glanduliflora* was the commoner of the two alpine heather species, also found in openings. Table 2.2 lists the notable species of this upper montane to treeline vegetation as of July, 1979.

Above the ragged treeline, about 1 km (0.6 mile) above Timberline Camp at approximately 1372 m (4500 ft) elevation, three distinct habitats were noted. The treeless north-trending ridges covered with stabilized pyroclastic deposits supported a meager flora of *Penstemon cardwellii*, *Achillea millefolium*, *Lupinus lepidus* var. *lobbii*, and *Eriogonum pyrolaefolium* var. *coryphaeum* as dominants (Fig. 2.1). A second habitat on either side of the ridges was a highly unstable pumice scree (Fig. 2.2); the only plant gaining a foothold here was *E. pyrolaefolium* var. *coryphaeum*, with its long trailing caudex anchoring the little rosettes from several centimeters upslope. The third habitat occurred on moister swales or benches at this same general elevation. Vegetation here formed a more dense cover, composed of *Luetkea pectinata*, *Fragaria virginiana*, *Antennaria microphylla*, and *Lycopodium sitchense*. Plants

FIG. 2.1 North slope of Mount St. Helens, with Dog's Head on left skyline. Note ridges of pyroclastic deposits and light-colored slopes of pumice; in the foreground is xeric grass–sedge–forb meadow (taken July 15, 1960).

FIG. 2.2   A view from the north slope of Mount St. Helens looking down-slope to Timberline Camp area. Pyroclastic deposits are in the foreground and whitish pumice screes elsewhere above timberline (taken July 15, 1960).

of this assemblage included two species of *Botrychium*, *Gaultheria ovati-folia*, and *G. humifusa*, *Carex* ssp., and *Saxifraga tolmiei*. Table 2.2 lists the notable plants of the treeless habitats. Although the Washington Native Plant Society's field party of July 1979 did not venture higher, a depauperate flora was noted at higher alpine areas by St. John (1976), who wrote, "The plants growing highest on the mountain are Pyrola-leaved Eriogonum (*Eriogonum pyrolaefolium* var. *coryphaeum*), Parry's Rush (*Juncus parryi*), Pussy Paws (*Spraguea umbellata*), Alpine Collomia (*Collomia debilis* var. *larsenii*) and Alpine Saxifrage (*Saxifraga tolmiei*)." St. John further remarked that "the summit (2950 m) is bare of vegeta-tion." It presumably still is, even though the summit is now 400 m (1300 ft) lower.

Our reconnaissance of the Mount St. Helens flora and vegetation also included a traverse of the remarkable Plains of Abraham, from north to south (Windy Pass to the head of Ape Canyon), which is a stretch of about 3 km (1.9 miles) of pumice flats. Forested and treeless stretches formed a mosaic without any appreciable change in elevation or topography. The forested sector of the Plains of Abraham consisted of subalpine fir and *Lupinus latifolius*, with a lesser number of lodgepole pine. Toward the lofty east base of the mountain proper, the forest gave way to a tree–scrub habitat composed of the same species. Stunted and deformed trees were common here because of repeated snow avalanche attack. Besides the common herbs of the timberline area, the Plains of Abraham supported other species common to pumice in the Cascades, notably *Polygonum newberryi*, *Phlox diffusa*, and *Aster ledophyl-lus*.

CONCLUSIONS

Two lasting impressions of the plant life prior to 1980 on Mount St. Helens (which could be called the Fujiyama of North America) have formed from a long association with the mountain and several of its volcanic neighbors in the Pacific Northwest. First, the upper montane vegetation of Mount St. Helens appeared to be xeric, sparse, and depauperate. The ragged, ill-defined timberline with its atypical assemblage of conifers and subordinate shrub–herb cover of low density and low species diversity seemed to be the result of two factors: the youthful age of the terrain and the composition of the substrate. The pumice as parent material had barely begun to form a soil profile since the last eruption of the nineteenth century. The more unstable pumice slopes were undoubtedly xeric during the short growing season.

The second impression comes from comparisons with other floras on Pacific Northwest volcanoes. The notable lack on Mount St. Helens of those numerous species inhabiting nearby volcanoes was symptomatic of the youthfulness and physical nature of the upper montane terrain. These observed deficiencies in Mount St. Helens flora are based only on a reconnaissance of the north and northeast slopes; species missing here may have been on other slopes or exposures of the mountain. Table 2.3 compiles the more significant deficiencies in the upper montane flora.

Impermanence, a key factor for most biotas, is accentuated in regions of active volcanism. The repeated eruptions of Mount St. Helens should remind us of what may be in store for most of the Pacific Northwest volcanoes. As they too resume activity, their floras and faunas will suffer decimation only to rebound with rapid and tenacious resilience.

LITERATURE CITED

APPLEGATE, E. I. 1939. Plants of Crater Lake National Park. *Amer. Midland Natur.* 22:225–314.

CARLQUIST, S. 1970. *Hawaii, A Natural History.* Natural History Press, Garden City, New York. 463 pp.

COOKE, W. B. 1940. Flora of Mt. Shasta. *Amer. Midland Natur.* 23:497–572.

———. 1962. On the flora of the Cascade Mountains. *Wasmann J. Biol.* 20:1–68.

FRANKLIN, J. F., and C. T. DYRNESS. 1973. Natural Vegetation of Oregon and Washington. *USDA Forest Service Gen. Tech. Rep. PNW-8.*

FRANKLIN, J. F., and C. WIBERG. 1979. Goat Marsh research natural area. *Supplement No. 10 to Federal Research Natural Areas in Oregon and Washington: A Guidebook for Scientists and Educators, 1972.* USDA Forest Service, Pacific Northwest Forest and Range Experiment Station.

FRANKLIN, J. F., F. C. HALL, C. T. DYRNESS, and C. MASER. 1972. Cedar Flats research natural area. *Federal Research Natural Areas in Oregon and Washington: A Guidebook for Scientists and Educators.* USDA Forest Service, Pacific Northwest Forest and Range Experiment Station.

GILLETT, G. W., J. T. HOWELL, and H. LESCHKE. 1961. A flora of Lassen Volcanic National Park, California. *Wasmann J. Biol.* 19:1–185.

HARRIS, S. L. 1980. *Fire and Ice: The Cascade Volcanoes.* The Mountaineers and Pacific Search Press. Seattle, Washington. 316 pp.

HITCHCOCK, C. L., and A. CRONQUIST. 1973. *Flora of the Pacific Northwest.* Univ. of Washington Press, Seattle, Washington. 730 pp.

IRELAND, O. L. 1968. Plants of the Three Sisters region, Oregon Cascade Range. *Mus. Nat. Hist. Univ. of Oregon Bull.* 12:1–130.

JONES, G. N. 1938. The flowering plants and ferns of Mount Rainier. *Univ. of Washington Publ. in Biol.* 7:1–192.

KAWANO, S. 1971. Studies on the alpine flora of Hokkaido, Japan. I. Phytogeography. *J. Coll. of Liberal Arts, Toyama Univ., Japan, Pt. Nat. Sci.* 4:13–98.

KRUCKEBERG, A. R. 1969a. Soil diversity and the distribution of plants, with examples from western North America. *Madrono* 20:139–154.

———. 1969b. Plant life on serpentinite and other ferromagnesian rocks in north-western North America. *Syesis* 2:16–114.

———.1979. Mt. St. Helens Field Trip. *Douglasia* 3(3):9–10.

LAWRENCE, D. G. 1954. Diagrammatic history of the northeast slope of Mt. St. Helens, Washington. *Mazama* 36:41–44.

LIND, E. M., and M. E. W. MORRISON. 1974. *East African Vegetation*. Longman Group Ltd., London. 257 pp.

LITTLE, E. O., Jr. 1941. Alpine flora of San Francisco Mountain, Arizona. *Madrono* 6:65–81.

MUNZ, P. A., and D. D. KECK. 1959. *A California Flora*. Univ. of California Press, Berkeley and Los Angeles. 1,681 pp.

NUMATA, M. (ed.). 1974. *The Flora and Vegetation of Japan*. Kodansha Limited, Tokyo, and Elsevier Scientific Publishing Co., New York. 294 pp.

PECK, M. E. 1961. *A Manual of the Higher Plants of Oregon*. 2d ed. Binfords and Mort, Portland, Oregon. 936 pp.

SMATHERS, G. A., and D. MUELLER-DOMBOIS. 1974. *Invasion and Recovery of Vegetation after a Volcanic Eruption in Hawaii*. National Park Service Sci. Monog. Ser. No. 5. Govt. Printing Office, Washington, D.C. 129 pp.

ST. JOHN, H. 1976. The Flora of Mt. St. Helens, Washington. *The Mountaineer* 70(7): 65–77.

ST. JOHN, H. and E. HARDIN. 1929. Flora of Mt. Baker. *Mazama* 11:52–102.

UGOLINI, F. C., and R. J. ZAZOSKI. 1979. 4. Soils derived from tephra. *In* P. I. Sheets, and D. K. Grayson, eds. *Volcanic activity and human ecology*, pp. 83–124. Academic Press, New York.

WIGGINS, I. L., and D. M. PORTER. 1971. *Flora of the Galapagos Islands*. Stanford Univ. Press, Stanford, California. 998 pp.

TABLE 2.1
Some Plants of Pacific Northwest Volcanoes[a]

Taxon	Status[b]	Range[c]
Agastache parvifolia Eastw.	I	Lassen; other n. Calif. lavas
Allium tribracteatum Torr.	I	Shasta (to s. Calif.)
Anemone occidentalis Wats.	I	Lassen to Baker
Arabis platysperma Gray	I	Lassen to Hood; south to Sierras
A. rectissima Greene	I	S. Oregon
Arenaria pumicola Cov. & Leib.	E	Mazama
Arnica viscosa Gray	E	Mazama, Shasta
Aster gormanii (Piper) Blake	I	Jefferson
Botrychium pumicola Cov.	E	Mazama, Shasta
Campanula wilkinsiana Greene	I	Shasta; Trinity Co., Calif.
Carex straminiformis Bailey	I	Lassen, Shasta, and eastward
Castilleja cryptantha Pennell & Jones	E	Rainier
C. lemmonii Gray	I	Lassen (=*C. lassenensis*); Sierras
C. payneae Eastw.	I	Lassen to Cascades of Oregon
Collomia debilis (S. Wats.) Greene var. *larsenii* (Gray) Brand.	I or E	Rainier to Lassen
C. mazama Cov.	E	Mazama or McLoughlin
Corydalis caseana Gray	I?	Sierras; Shasta, Lassen
Cryptantha subretusa Jtn.	I?	Lassen; n. Calif., Oregon
Draba aureola Wats.	E	Lassen to Rainier
D. breweri Wats.	I?	Sierras; Shasta, Lassen
Erigeron aureus Greene	I?	Wash., s. Brit. Columbia
E. elegantulus Greene	I?	Lassen; s. and e. Oregon
E. lassenianus Greene	I	Lassen; Sierras
Eriogonum marifolium T. & G.	I?	Sierras; volcanic Cascades
E. ochrocephalum Wats.	I?	Sierras; Lassen
E. pyrolaefolium Hook. var. *coryphaeum* T. & G.	I	Lassen to Rainier
E. ursinum Wats.	I?	Sierras; Lassen
Erythronium klamathense Appleg.	I?	S. Cascades to Calif.
Fritillaria adamantina Peck	E?	Diamond Lake
Holodiscus microphyllus Rydb. var. *glabrescens* (Greenm.) Ley.	I	Lassen; n. Calif.
Hulsea nana Gray	E	Lassen to Rainier
Lesquerella occidentalis Wats.	I?	N. Calif., Oregon, Nevada; Lassen
Lilium washingtonianum Kell.	I	Hood to Calif.

TABLE 2.1
SOME PLANTS OF PACIFIC NORTHWEST VOLCANOES[a] (continued)

Taxon	Status[b]	Range[c]
Luina (Rainiera) stricta (Greene) Rob.	I	Hood to Rainier
L. albicaulis Dougl. ex Hook var. *shastensis* (Heller) C. P. Sm.	I	Sierras to s. Oregon
L. andersonii var. *christinae* (Heller) Munz	E?	Lassen, (Shasta?), Crater Lake
Lupinus lepidus Dougl. var. *lobbii* (Gray) Hitch.	I	Rainier to Lassen
L. obtusilobus Heller	I?	Sierras; Lassen
Machaeranthera shastensis Gray	I	Shasta to central Oregon
Panicum thermale Bol.	I	Lassen; w. North America (hot springs)
Pedicularis ornithorynchya Benth.	I?	Rainier to s. Alaska
P. rainierensis Pennell & Warren	E	Rainier
Penstemon cinereus Piper	I	N. Calif.; Oregon
P. cinicola Keck	E?	Crater Lake to Jefferson
P. shastensis Keck	I	N. Calif.
P. neotericus Keck	I	Lassen; n. Sierras
Phlox hendersonii (E. Nels.) Cronq.	I	Washington to Hood
P. bryoides Nutt.	I?	Lassen; Great Basin, Rocky Mtns.
Polygonum davisiae Brew. ex Gray	I?	Sierras; Lassen; s. Oregon
P. newberryi Small	I	Lassen to Rainier
P. shastense Brew. ex Gray	I?	Sierras; Lassen, Shasta; s. Oregon
Polemonium elegans Greene	E?	Rainier to s. British Columbia
Raillardella argentea (Gray) Gray	I	Sierras; volcanic Cascades to Three Sisters
Ranunculus gormanii Greene	I	Mazama to Three Sisters
Ribes erythrocarpum Cov. & Lieb.	E?	S. Oregon to Crater Lake
Saxifraga tolmiei T. & G.	I	Sierra Nevada to Alaska
Sedum oregonense (Wats.) Peck	I	S. Oregon
Silene montana Wats. ssp. *montana*	I?	Sierras to Crater Lake
S. suksdorfii Robins.	I	Lassen to Rainier
Smelowskia ovalis Jones	E	Three Sisters to Rainier
S. ovalis Jones var. *congesta* Roll.	E	Lassen
Streptanthus tortuosus Kell var. *orbiculatus* (Greene) Hall	I?	Sierras; Lassen, Shasta

TABLE 2.1
SOME PLANTS OF PACIFIC NORTHWEST VOLCANOES[a] (continued)

Taxon	Status[b]	Range[c]
Tauschia stricklandii (Coult. & Rose) Math. & Const.	E	Rainier
Viola purpurea Kell. ssp. *dimorpha* Baker & Clausen	I?	Lassen; Sierras; central Oregon

[a]Compiled from Applegate (1939), Cooke (1940, 1962), Gillett et al. (1961), Ireland (1968), Jones (1938), and St. John and Hardin (1929). Nomenclature based on Hitchcock and Cronquist (1973), Munz and Keck (1959), and Peck (1961).

[b]E = endemics, I = indicator, and ? = doubtful status.

[c]Place name abbreviations: California: Lassen = Mount Lassen, Shasta = Mount Shasta, Sierras = Sierra Nevada. Oregon: Mazama = Mount Mazama (Crater Lake area), Jefferson = Mount Jefferson, Hood = Mount Hood. Washington: Adams = Mount Adams, Rainier = Mount Rainier, Baker = Mount Baker.

TABLE 2.2
SOME NOTABLE PLANTS OF THE UPPER MONTANE, MOUNT ST. HELENS, SUMMER 1979[a]

Fern and Fern Allies

Athyrium filix-femina (L.) Roth. +
Botrychium spp. ++
Cryptogramma crispa (L.) R.Br., rare
Lycopodium sitchense Rupr. ++
Polystichum munitum (Kaulf.) Presl., rare

Conifers

Abies amabilis (Dougl.) Forbes +
A. lasiocarpa (Hook.) Nutt. ++
A. procera Rehder ++
Juniperus communis L., rare
Pinus contorta Dougl. ++++
P. monticola Dougl. +
Pseudotsuga menziesii (Mirbel) Franco +
Tsuga heterophylla (Raf.) Sarg. +
T. mertensiana (Bong.) Carr. ++

Monocots

Agrostis diegoensis Vasey ++++
Carex spp. +++
Danthonia spp. ++
Festuca viridula Vasey +
Juncus parryi Engelm. ++
Luzula spp. ++
Phleum alpinum L. ++
Sitanion jubatum Smith +
Smilacina racemosa (L.) Desf., local
S. stellata (L.) Desf., local
Spiranthes romanzoffiana Cham. ++
Trillium ovatum Pursh, local
Veratrum viride Ait., rare

Dicots

Alnus sinuata (Regel) Rydb. +
Antennaria lanata (Hook.) Greene +
A. microphylla Rydb. ++

TABLE 2.2
SOME NOTABLE PLANTS OF THE UPPER MONTANE,
MOUNT ST. HELENS, SUMMER 1979[a] (continued)

Dicots (cont'd.)

Anaphalis margaritacea (L.) B. & H. +++

Achillea millefolium L. ssp. *lanulosa* (Nutt.) Piper ++

Arctostaphylos nevadensis Gray +

A. uva-ursi (L.) Spreng. +

Arnica latifolia Bong. +++ (woods)

Aster ledophyllus Gray ++

Castilleja sp. ++

Epilobium angustifolium L. ++

Eriogonum pyrolaefolium var. *corypheum* T. & G. +++

Eriophyllum lanatum (Pursh) Forbes +

Fragaria virginiana Duchesne +++

Gaultheria humifusa (Grah.) Rydb. +

G. ovatifolia Gray +

Heuchera micrantha Dougl., local

Hieracium albiflorum Hook. ++

H. gracile Hook. ++

H. hybrid (*albiflorum* X *gracile*), rare

Luetkea pectinata (Pursh) Kunze ++++

Lomatium martindalei Coult. & Rose +

Lupinus latifolius Agardh. ssp. *subalpinus* +++ (open woods)

L. lepidus Dougl. ssp. *lobbii* (Gray) Hitch. ++++ (pumice)

Luina hypoleuca Benth., rare

Menziesia ferruginea Smith, rare

Microseris alpestris (Gray) Q. Jones (?)

Montia sibirica (L.) Howell, local

Phacelia leptosepala Rydb., local

Phlox diffusa Benth. ++

Phyllodoce empetriformis (Sm.) D. Don ++

P. glanduliflora (Hook.) Cov. +++

Populus trichocarpa T. & G. +

Penstemon cardwellii Howell ++++

Polygonum newberryi Small +++

P. minimum Wats., rare

Pyrola asarifolia Michx., local

P. secunda L., rare

Rubus lasiococcus Gray, local

Saxifraga tolmiei T. & G. +

Salix spp. +++ (water courses)

Sibbaldia procumbens L. +

Sorbus scopulina Greene +++

S. sitchensis Roemer ++

Spiraea densiflora Nutt., local

Spraguea umbellata Torr. ++

Trautvetteria caroliniensis (Walt.) Vail, local

[a]Observations made at Timberline Camp area and on the Plains of Abraham, July 27–29, 1979. Estimates of frequency: rare, local, + = infrequent, ++ = frequent, +++ = common, ++++ = abundant.

TABLE 2.3

SOME CASCADIAN SUBALPINE AND ALPINE SPECIES EXPECTED BUT
NOT PRESENT ON MOUNT ST. HELENS (PREERUPTION)

Ferns and fern allies

Cheilanthes gracillima D. C. Eat.
Polypodium hesperium Maxon
P. kruckebergii Wagner
Woodsia spp.

Liliaceae

Allium cernuum Roth
Erythronium montanum Wats.[a]
Fritillaria lanceolata Pursh
Lloydia serotina (L.) Sweet.

Woody dicots

Salix nivalis Hook.
S. cascadensis Cockerell
Potentilla fruticosa L.
Kalmia microphylla (Hook.) Heller
Cassiope mertensiana (Bong.) G. Don
Vaccinium deliciosum Piper

Conifers

Chamaecyparis nootkatensis (D. Don)
   Spach[a]

Grasses and graminoids

Agropyron spp.
Calamagrostis spp.
Deschampsia atropurpurea (Wahl.)
   Scheele

Orchidaceae

Calypso bulbosa (L.) Oakes
Habenaria dilatata (Pursh) Hook.

Herbaceous dicots: Polypetalae

Aconitum columbianum Nutt.
Anemone occidentalis Wats.
A. drummondii Wats.

Herbaceous dicots: Polypetalae
(cont'd.)

Arabis spp.
Arenaria spp.
Dryas octopetala L.
Draba aureola Wats.
Draba (other perennial spp.)
Epilobium latifolium L.
E. luteum Pursh
Lathyrus spp.
Leptarrhena pyrolifolia (D. Don)
   R. Br.
Ligusticum spp.
Oxyria digyna (L.) Hill.
Parnassia fimbriata Konig.
Potentilla diversifolia Lehm.
P. flabellifolia Hook.
Ranunculus escholtzii Schlect.
R. suksdorfii Gray
Rubus pedatus J. E. Smith
Saxifraga bronchialis L.
S. ferruginea Grah.
S. oppositifolia L.
Sedum lanceolatum Torr. var.
   rupicolum (Jones) Hitch.
S. divergens Wats.
Silene acaulis L.
S. douglasii Hook.
S. parryi (Wats.) Hitch. & Mag.
S. suksdorfii Robins.
Smelowskia ovalis Jones
Spiraea betulifolia Pall. var. *lucida*
   (Dougl.) Hitch.
Thalictrum spp.
Thlaspi fendleri Gray

Herbaceous dicots: Gamopetalae

Aster alpigenus (T. & G.) Gray[a]
Artemisia norvegica Fries
Chimaphila umbellata (L.) Bart.
Cirsium edule Nutt.

TABLE 2.3
Some Cascadian Subalpine and Alpine Species Expected but Not Present on Mount St. Helens (Preeruption) (continued)

Herbaceous dicots: Gamopetalae
(cont'd)

Douglasia laevigata Gray
Empetrum nigrum L.
Erigeron peregrinus (Pursh) Greene[a]
E. compositus Pursh
E. aureus Greene
Gentiana calycosa Griseb.
Haplopappus lyallii Gray
Hulsea nana Gray
Mimulus lewisii Pursh[a]
Pedicularis contorta Benth.
P. bracteosa Benth.

P. groenlandica Retz.
P. ornithoryncha Benth.
Phacelia sericea (Grah.) Gray
Polemonium elegans Greene
P. pulcherrimum Hook.
Saussurea americana Eat.
Senecio triangularis Hook.
Solidago multiradiata Hit. var.
    scopulorum Gray
Veronica cusickii Gray

[a]Species listed by St. John (1974) but not seen by author.

3

Monitoring Effects of the Mount St. Helens Eruptions on the Toutle River Drainage Basin

Elizabeth M. W. Pincha

ABSTRACT

Aerial and ground photographic reconnaissance performed after the May 18, 1980, eruption of Mount St. Helens have served as valuable tools for monitoring the cyclic rainfall and runoff erosional processes active in the Toutle River drainage basin. These data, along with other field data, have enabled evaluation and assessment of the devastated areas and have prompted human efforts to reduce further destruction. The photographic records also provided a basis for recommendations concerning volcanic and fluvial geomorphic changes occurring in the Toutle River system in response to storm and volcanic activities. Land managers must consider the following: restriction on future tree salvaging, continued dredging of the Debris Restraining Structure (DRS) N-1 reservoir, repair of the DRS N-1 structural breaches, construction of holding or settling ponds in the upper basin above DRS N-1, and control of the draining of Spirit Lake. Successful short-term efforts have been made in dredging throughout the drainage basin and beyond, evaluating the structural soundness of the lakes impounded by volcanic debris, controlling the drainage of ponds and lakes, seeding and fertilizing denuded areas, restraining massive volumes of bed load sediment, and identifying and measuring the hydrologic characteristics of the blast zone. A lack of funds, however, has halted some of these efforts. Funds are needed to repair the DRS N-1 structural breaches and to dredge its reservoir and the lower Toutle and Cowlitz rivers. Without further efforts and funding to restrain the unstabilized sediment movement within the Toutle River drainage basin and surrounding areas, a potential threat to life and property equal to or greater than the mudflows of May 18–19, 1980, exists in the upper basin.

INTRODUCTION

Aerial photographic reconnaissance flights made possible the monitoring, gathering of data, and subsequent analyzing of important geomorphic changes that occurred in the unique blast area and mudflow deposits within the Toutle River drainage following the major eruption of the Mount St. Helens volcano on May 18, 1980. A considerable amount of data has been gathered by many investigators using a great variety of methods. An independent research of otherwise inaccessible devastated zones, however, was achieved through photographic flights to record geomorphic changes directly influenced by volcanic and hydrologic cycles, and by human efforts to exist within these cycles. Periodic aerial and supplemental ground photographic trips were made to assess and identify those geomorphic alterations in the Toutle River drainage basin that resulted from rainfall and subsequent surface runoff erosion and from volcanic mudflows. Geomorphic and fluvial geomorphic changes monitored and evaluated include hillslope stability and morphology, channel morphology, sediment type, sediment movement and yield, and drainage pattern development.

The Toutle River drainage empties into the Cowlitz River, which in turn flows into the Columbia River. Since the Columbia River is an avenue of world trade shipping, any devastation within the Toutle drainage will significantly affect these navigation lanes. Thus, any future volcanic activity in the Toutle River drainage becomes important to this relationship. It is necessary, therefore, to repeatedly monitor and evaluate changes in volcanic activity so that possible damaging effects on the drainage basin can be predicted and action can be taken.

The Toutle River drainage basin is shown in Figure 3.1, including the devastated area north of the breached caldera, the entire upper North and South Forks, the main Toutle channels, and the Green River drainage. This study also includes the Cowlitz River past its confluence with the Toutle, and the Columbia River upstream above the Kalama and downstream from the Cowlitz. Figure 3.2 shows channel cross-sections for locations indicated on Figure 3.1.

CHRONOLOGICAL EVENTS OF THE MAY 18 ERUPTIONS

MOUNT ST. HELENS VOLCANIC ACTION

The Cascade range consists of several north–south trending stratovolcanoes formed in the North American plate over the eastern edge of the subducting Juan de Fuca plate (Bingham 1980, Decker and

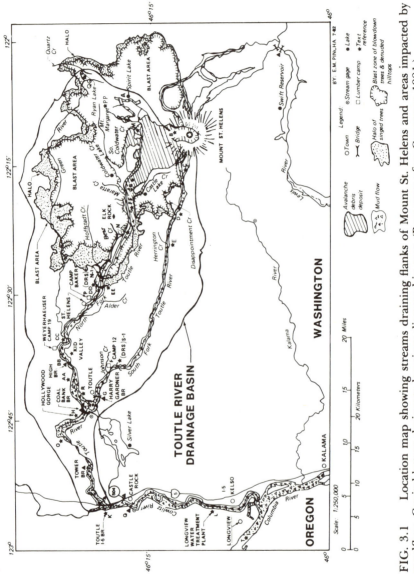

FIG. 3.1 Location map showing streams draining flanks of Mount St. Helens and areas impacted by mudflow. Capital letters designate locations discussed in text. (Base map after Cummans 1981.)

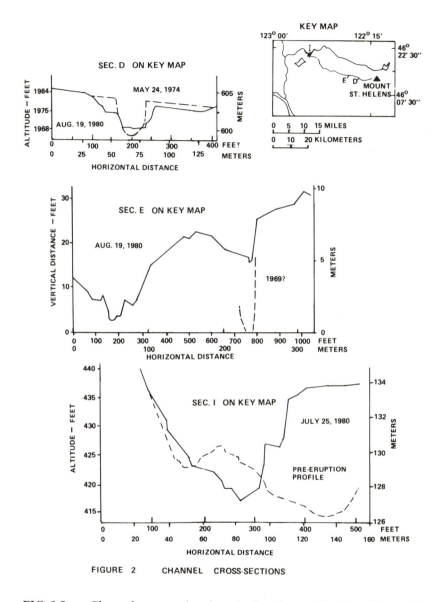

FIGURE   2   CHANNEL   CROSS-SECTIONS

FIG. 3.2   Channel cross-sections keyed to locations on Fig. 3.1. (After Janda et al. 1981 and Lombard et al. 1981.)

Decker 1981, Rau 1980). One of these stratovolcanoes is Mount St. Helens, which is located over the eastern edge of this plate in a weakened zone between two side-slipping transverse faults, trending northwest to southeast. Many of these volcanoes were active throughout the nineteenth century. In the beginning of the twentieth century, Mount Lassen in California was active as early as 1914 with only a single eruptive dacitic lava flow on May 19, 1915 (Loomis 1926).

After 123 years of quiescence (Hoblitt et al. 1980), Mount St. Helens resumed activity on March 20, 1980, with earthquakes of magnitudes greater than 4 on the Richter scale. Near its summit on the north flank, by May 17 a noticeable graben appeared as a 97.5 m (320 ft) bulge named the Forsyth Bulge. This proved to be a critically unstable multifractured land mass, sensitive to the slightest earth disturbance (Moore and Albee 1981). At 8:32 A.M. on May 18, 1980, the triggering disturbance came as an earthquake of magnitude 5 (Mullineaux and Crandell 1981), causing one of the largest rock avalanche and subsequent landslides in historical times (Decker and Decker 1981). Two geologists, Keith and Dorothy Stoffel, flying in a light plane directly over the summit, captured this event on film (Korosec et al. 1980). From their aerial photographs, photographs taken by ground observers, and computer graphic analyses, a logical and realistic scenario of events is proposed (Korosec et al. 1980, Moore and Albee 1981, Rosenfeld 1981). Three distinct blocks of the north flank of Mount St. Helens had rotationally slumped along a lubricated deep-seated slip plane. The resultant caldera formed the largest breached structure identified in modern times. The slumped material surged into Spirit Lake basin (Fig. 3.1, location A) displacing waters as a destructive wave east, northeast, and north toward Mount Margaret (Fig. 3.1, location PP).

The ensuing nuée ardente ("hot glowing cloud") of intensely heated andesitic flank fragments, highly charged with gas and superheated steam, leveled 391 km^2 (96,000 acres, or 151 sq miles) (Pruitt et al. 1980) of forest in the Gifford Pinchot National Forest, along with forest on private and state lands. Immediately north of the caldera, in a 120° fan-shaped lobe, trees and soil were obliterated to expose bedrock for a radius of 13 km (8 miles). Outward from this zone for another 5–8 km (3–5 miles), trees were stripped, snapped, uprooted, or blown down in place. The trees felled by the blast force conformed to radial lines directed away from the caldera for the first 13 km (8 miles). The position of the downed trees, mapped from photographs (Kieffer 1981), led to an inference that the dense multiphased cloud had moved as a single unit in a very complex manner. This heavily loaded cloud of solid, liquid, and gaseous particles, following the laws of gravity, conformed to the topography as a ground-hugging mass. Beyond the initial 13 km (8 mile) zone, however, the random and jumbled disarray of downed trees, as seen in the aerial photographs, suggested that another force or forces caused the formation of new eddies and turbulences as convective updrafts. This force might have resulted when a second blast wave from the erupting mountain interacted with the initial blast and moved at greater than sonic speeds, reacting with the local landforms. Evidence for this hypothesis is that the mapped pattern shows blown down trees in some valleys lying crossways with respect to the downhill slope or back toward the mountain (Kieffer 1981). Toward the edge of the blast zone, the abrasively charged and intensely heated load became diminished in content and heat. The remaining gases still retained sufficient heat to scorch trees and enough energy to surge outward as balls, eddies, and fingers to burn and destroy vegetation. In a halo radiating from the caldera, 4 km (2.5 miles) wide and 29 km (18 miles) long,[1] erratic patterns of blown down and singed trees were formed (Fig. 3.1). The temperature range of the singed tree needles was from 50 to 250°C (120 to 480°F), as determined later by tests (Winner and Casadewell 1981). The lethal temperature for fir leaves is 48°C (117°F).

After 1 yr, substantial revegetation had occurred on the west- and north-facing slopes located within the devastated 13 km (8 mile) zone and north of the caldera. These slopes had been protected by an insulating layer of snow during the eruption. In contrast, the top third of the south-facing slopes had been bared to bedrock and no vegetation had yet returned.

ROCK AVALANCHE AND PYROCLASTIC FLOW

When the Forsyth Bulge failed, mammoth rock and snow avalanches were released toward the north (Fig. 3.1, location B) (Mullineaux and Crandell 1981). The changing morphology of the volcanic ramp as of October 1980 and May 1982 is shown in Figures 3.3 and 3.4. Fragments of cryptodome, glacial ice blocks, snow, and flank rocks were propelled 100–110 m/sec (220–245 mph) (Kieffer 1981) down onto the upper valley of the North Fork Toutle. A triangular wedge of the avalanche was about 122 m (400 ft) deep near the volcano, but its final length was obscured by subsequent pyroclastic deposits. Many creeks flowing into the North Fork Toutle were impounded by this and later debris (Fig. 3.5). The avalanche displaced the water from Spirit Lake and filled its basin with 46 m (150 ft) of material (Cummans 1981). The

FIG. 3.3 Mount St. Helens volcanic ramp (10/28/80). Pyroclastic and debris flows of five major eruptions. North Fork Toutle at right of picture. Altitude 2120 m (7000 ft), view looking west. Location B on Fig. 3.1.

FIG. 3.4 Mount St. Helens volcanic ramp (5/23/82). Pyroclastic-filled ramp with subsequent erosive head cutting occurring in debris. Note fresh ashfall from 4 hr previous eruption; note steam from the new dome. Altitude 1970 m (6500 ft), view looking southwest. Location B on Fig. 3.1.

FIG. 3.5 North Fork Toutle River valley (2/4/81). Hummocky block pyroclastic debris flow, 15–45 m (50–150 ft) high, 107 m (350 ft) deep (lower left corner); Coldwater Ridge (bottom center); Coldwater lake impoundment dam (right center). Altitude 2120 m (7000 ft), view looking west. Location GG on Fig. 3.1.

FIG. 3.6 Spirit Lake and breached Mount St. Helens (10/28/80). Log raft fills 85% of surface area of Spirit Lake; denuded slopes form dendritic drainage pattern; erosive head cutting occurring in volcanic ramp; upper reaches of North Fork Toutle relatively smooth. Altitude 2120 m (7000 ft), view looking south–southwest. Location PP on Fig. 3.1.

FIG. 3.7   Spirit Lake and basin (10/28/80). Hummocky
block deposits at north end of lake, 15−45 m (50−150 ft)
high; log raft covering 85% of water surface; dendritic
drainage pattern on hillslope at lower right. Altitude 3300 m
(10,000 ft), view looking north. Location A on Fig. 3.1.

FIG. 3.8   Spirit Lake, basin, and enclosing hills (3/5/82).
Water level this date higher than level of 10/28/80; log raft
smaller, possibly frozen together; Mount Rainier in background. Altitude 2120 m (7000 ft), view looking north.
Location A on Fig. 3.1.

displaced surge of water rapidly ravaged the surrounding forest, de-limbing and uprooting 300-yr-old Douglas fir trees (Pruitt et al. 1980). It brought these fallen logs back into the lake basin where they collected into a massive raft (Figs. 3.6 and 3.7). Years later, this raft continues drifting to and fro at the whim of the prevailing winds (Fig. 3.8).

Upon release of the structural cap (the Forsyth Bulge), the nuée ardente contained the remaining pulverized dacite and andesite rock of Mount St. Helen's north flank. Combined with this ungraded rock was ground water charged with superheated steam and gases. This lubricated viscous mass slid and rolled at velocities greater than 300 m/sec (670 mph) (Kieffer 1981, Moore and Sisson 1981) northward into the river valley, depositing a 30 m (100 ft) layer over the avalanche debris and a 15 m (50 ft) layer into Spirit Lake (Cummans 1981). Moving due north, it crested a 352 m (1156 ft) ridge separating Spirit Lake and South Coldwater Creek (Fig. 3.1, location M). The distance from the ridge to the debris terminus is 12.2 km (7.6 miles), with a depth ranging from 15 to 46 m (50 to 150 ft). Toward the west, the nuée ardente debris wedge created a wide destructive path that changed the geomorphology of the upper reaches of the North Fork Toutle. The thickest part of the wedge was located southwest and west of Spirit Lake at a depth of 106.7 m (400 ft) with its distal edge located 22.9 km (14.2 miles) downstream (Fig. 3.1, location N). Thin tails extended as far as 39 km (24 miles) west beyond the confluences of Bear and Hoffstadt creeks. Northeast of the caldera, hummocky pyroclastics and debris deposits from 15.2 to 45.7 m (50 to 150 ft) deep were deposited on the northern lobes of Spirit Lake (Fig. 3.7), providing positive evidence of the avalanche surge (Cummans 1981, Korosec 1981, Moore and Sisson 1981, Watt 1981).

SOUTH FORK TOUTLE MUDFLOW

Within 20 min of the major eruption, hot pyroclastic and liquid debris from the volcano had melted the blocks of glacial ice and snow ejected into the headwaters of the South Fork Toutle (Fig. 3.1, location C) (Cummans 1981). This meltwater and unsorted debris surged at velocities of 64 km/hr (40 mph). Rushing downstream, the overbank surges at the river bends approached 160 km/hr (100 mph). It quickly reshaped the morphology of the valley by scouring and filling the channels, collapsing banks, and ripping out riparian trees along with their root wads. As the discharge increased, its velocity decreased to less than 40 km/hr (25 mph) approaching Weyerhaeuser's storage camp 12 (Cummans 1981) (Fig. 3.1, location F). The thick flow swept stockpiled logs as large as 1.5 m (5 ft) in diameter into the mixture and chaotically distributed huge logging equipment and trucks among the knocked-down riparian alders and bigleaf maples. This dynamic discharge of large particles, debris, and mud slurry eroded banks and scoured the upper reaches of the South Fork Toutle, forming new channels. The sediment deposited on new flood plains, shown in Figure 3.1 (location E) and outlined on the channel cross-sections in Figure 3.2 (locations D and E). The poorly sorted discharge continued downstream as a 3.7 m (12 ft) flash flood, destroying or badly damaging highway and logging roads and 27 bridges (Lombard et al. 1981, Schuster 1980), but leaving the Harry Gardner Bridge intact on the South Fork Toutle (Fig. 3.1, location G).

Beyond the confluence of the North and South Fork Toutle, this swirling log-swelled, swiftly moving debris flow weakened the abutments of the SR 504 Coal Bank Bridge (Fig. 3.1, location I and Fig. 3.2, location I). Past the bridge, within the narrow Hollywood Gorge (Fig. 3.1, location H), a lake formed behind the logjam of debris, railroad trestles, logs, and uprooted trees; this lake reached a maximum depth of 16 m (53 ft). This depth was measured from the height of mudlines (maximum stage marks) on the bankside trees. The logjam broke at 11:51 A.M. The entrained log raft and debris quickly drained from the impoundment as it flooded westward toward the Cowlitz River at about 29 km/hr (18 mph). When this log-filled flood swept past the Toutle I-5 bridge into the deeper and slower flowing Cowlitz (Fig. 3.1, location K), larger particles from the bed load settled out and formed a delta. As the remaining discharge passed the Long-view Water Treatment Plant (Fig. 3.1, location L) at 4:15 P.M., the water temperature was 11°C (52°F) (Cummans 1981). Many excellent photographs, particularly from NASA U-2 flights, were taken of this log raft and debris flow. These photographs were used to identify the upstream sediment delta that developed at the mouth of the Toutle and the sediment lobe that formed downstream in the Cowlitz.

A second mudflow originated in the headwaters at 2:00 P.M. and advanced only to the lower reaches of the South Fork Toutle. This second flow contained material that was darker and different in texture as it overlaid the first flow on the South Fork's upper reaches. Much of this second layer has eroded and probably has mixed with other sediment in the Cowlitz or in the dredged spoils.

NORTH FORK TOUTLE MUDFLOW

In the upper reaches of the North Fork Toutle, meltwater from snow and 3–5 ton glacial ice blocks, combined with volcanically created rain, completely filled cavities in the hummocky debris deposits (Fig. 3.5).

At 1:04 P.M., a thick, dark brown, highly viscous fluid mass moved slowly down the narrow channel along the northern side (Cummans 1981). A second flow then overtopped depressions in the deposits on the southern side, scouring and laterally enlarging the channel into a wide valley. By 2:30 P.M., the northern flow had overrun and destroyed Weyerhaeuser's maintenance and main storage Camp Baker (Fig. 3.1, location P). As it continued its destructive path through riparian vegetation, structures, railroad tracks, and equipment, the mortarlike discharge grew into a huge and dangerous load. From a hilltop near Kid Valley overlooking a narrow gorge, a fully loaded and partially sunk logging truck was seen floating down the river. As the river basin widened, the front of the flow spread laterally over the banks (Fig. 3.9), and large unsorted debris such as bridge abutments, pieces of equipment, railroad ties, culverts, and warped rails (from Weyerhaeuser's Camp Baker) became distributed over the flood plains (Fig. 3.12). In addition, bed load sediment composed of pebble- to boulder-sized rocks, logs, and tree stumps and wads created new flood plains (Fig. 3.11). Some logs and trees were entrenched and angled 30° to the

FIG. 3.9   Location 1.6 km (1 mi.) west of confluence of Green and North Fork Toutle; Red Zone Road Block (3/5/82). Note straight low gradient river valley with many braided and meandering channels; sand and gravel bars; and embedded trees, logs, boulders. Aerial view of scenes shown in Figs. 3.10–3.12. Altitude 760 m (2500 ft), view looking southwest. Location CC on Fig. 3.1.

FIG. 3.10   North Fork Toutle, Red Zone (9/28/80).
Entrenched "bayonet" logs (sharp end, streamlined to flow)
in the~2 billion m³ (2.7 million yd³) volcanic debris mudflow
of 5/19/80. Note high stage mark on opposite bank tree
trunks. View looking northeast. Location CC on Fig. 3.1.

flow (Janda et al. 1981) (Fig. 3.10). Exposed bark and cambium layers
were either removed or polished to a silky finish by abrasive waters
filled with particles of volcanic glass, pumice, rhyolite, andesite, and
feldspar. As the viscous, thick slurry flowed around the logs, they
became ensnared and embedded in the river pavement near the Red
Zone Roadblock (March, 1982) and on SR 504 (Fig. 3.1, location CC;
Figs. 3.8–3.12). The logs were shaped and sharpened into typical
"bayonet" forms (Fig. 3.10). With each following rainstorm, the poorly
consolidated pavement of the river and its debris became subject to
extreme erosion. The material was then transferred into the downstream movement.

Many large boulders settled or became lodged between trees on
the flood plain. Houses caught in the flow were either lifted from their
foundations and moved with the mass or were coated and partially
filled with the thick, silty mud. A toppled Douglas fir was forced
directly into the front door of one house, now part of the new flood
plain. The advancing debris-widened flood turned a bend beyond the
confluence of the North and South Fork Toutle, and the weakened
decking of the 160 m (525 ft) Coal Bank Bridge snapped from its main
abutments (Schuster 1980) (Fig. 3.1, location I) and joined the already

FIG. 3.11   North Fork Toutle, Red Zone, looking upstream (12/27/80). Bankfull discharge from 12/25/80 storm fills entire channel, causing block failure of bank; stationary bed load creates standing waves that moved as bed load shifted. Compare with Figs. 3.9 and 3.12 for normal discharge. View looking south–southwest. Location CC on Fig. 3.1.

enormous load. Luckily, the velocity of the flow remained at about 8–11.3 km/hr (5–7 mph) (Cummans 1981), exerting minimal damage to the channels. The logjam of the South Fork Toutle had broken earlier, avoiding potentially destructive lake ponding.

At 12:15 P.M. on May 19, the flow reached the Cowlitz River near the town of Castle Rock (Fig. 3.1, location Q) (Cummans 1981). The dense sediment flow veered upstream into the Cowlitz—just as the mudflow from the South Fork Toutle had done—then turned downstream. With the sediment from this discharge, the channel capacity of the Cowlitz was reduced from its normal discharge of 2152 m³/sec (76,000 cfs) to 207 m³/sec (7300 cfs)[2] (Schuster 1980), which is a 96% decrease. The probable cause of the 4-km (2.5-mile) long upstream surge past the mouth of the Toutle River was a blocking delta formed earlier by large boulders, gravel particles, and debris of the bed load from the South Fork Toutle discharge, which rapidly settled in the reduced velocity of the Cowlitz (Fig. 3.1, location K). The onrushing North Fork Toutle sediment, joining the bed load and suspended load of the Toutle, encountered this delta and turned upstream into the

FIG. 3.12   North Fork Toutle, Red Zone (3/28/81). Block
failure of asphalt parking area because of bankfull discharge;
storm pipe, wood debris, and embedded logs are remnants of
bed load deposit; note people standing beyond single tree,
which is the tree in the flow in Fig. 3.11. Note high flood stage
line across trees on right. View looking south–southwest.
Location CC on Fig. 3.1.

slower moving Cowlitz, where many large particles settled from its bed
load. After a short period of time, the greater discharge of the Cowlitz
cleared the blocking delta and created a path for further downstream
advancement of the Toutle flow. Water temperature at Longview
Water Treatment Plant (Fig. 3.1, location L) rose to 32°C (92°F) (Cummans 1981). The sediment yield deposited into the Toutle, Cowlitz,
and Columbia rivers was 38 million m³ (50 million yd³) (see n.2). The
North Fork Toutle debris and mudflow action was equivalent to a
10,000-yr event.

ASHFALLS OF MAY 18 AND 19

Ash clouds erupting from Mount St. Helens on May 18 and 19 fell on
and near the mountain and drifted eastward. This ashfall layer varied
in thickness from 122 cm (48 in.) in and around the blast zone to less
than 1.2 cm (0.5 in.) near the Cowlitz River area (McLucas 1980).
During the fall of ash in those first two days, the two major mudflows
occurred, altering the blast zone. This combination provoked human
reaction.

HUMAN RESPONSES

COLUMBIA RIVER SHOAL RESTRICTIONS

On the Columbia River, a freighter captain alerted the U.S. Coast Guard that his ship went aground at 2:15 A.M. on May 19 (Fig 3.1, location T) (see n. 2). The second major debris flow and mudflow from the North and South Fork Toutle had reached the Columbia River navigation lanes. The normal Columbia navigation lane is a channel 12.2 m deep by 183 m wide (40 by 600 ft). The captain reported a raft of logs, debris, and dead fish more than 32 km (20 miles) long floating down the Columbia. The dead fish probably succumbed to the 32°C (90°F) temperature of the second mudflow in the Toutle and Cowlitz. A tolerable temperature for fish is less than 12.8°C (55°F). The U.S. Coast Guard confirmed that the sediment shoal had reduced the channel to a 4.3 m (14 ft) depth across the Columbia, a depth loss of 65%. The U.S. Army Corps of Engineers was informed of this development because the Corps is responsible for keeping these lanes open. The extent of this problem was fully appreciated when the Coast Guard warned captains of 50 ships anchored off the mouth of the Columbia not to proceed upstream and captains of 30 ships docked in ports at the cities of Portland and Longview not to move downstream.

DREDGING OPERATIONS AND FUTURE RECOMMENDATIONS

The Corps of Engineers quickly dispatched three hopper dredges into the area (see note 2). Within 5 days, the dredges had cleared 9.9 million m³ (13 million yd³) of debris through the 4.3 m (14 ft) shoal on the Oregon side (see note 1). This opened an emergency channel that permitted the passage of ships escorted by the Coast Guard, during each high tide "window." These ships required a minimum 10.7 m (35 ft) depth clearance. The last land-locked ship moved downriver on June 14, almost a month after the eruption. The Corps commissioned into action a pipeline dredge from the Port of Portland and five contract dredges to work along with the Corps' hopper dredges. This formidable task force restored the navigation lanes to full capacity by the end of November, after removing about 38 million m³ (50 million yd³) (see n. 2) of sediment. In addition to this route in the Columbia, the Corps restored 70% of the channel capacity of 1415 m³/sec (50,000 cfs) to the Cowlitz River.

    A serious dredging problem that required an immediate solution was the disposal of dredge spoils. As the hot, sediment-rich load of highly viscous, fine-grained, ash-filled slurry moved downstream in the Toutle River, it spilled over the banks, adhering to any and all surfaces and forming a clay, sand, and mud layer a maximum thickness of 7.6 cm (3 in.) (Janda et al. 1981). The mud-encased riparian trees (Fig. 3.10) were presumed dead or dying as a result of the effects of hot mud encircling their trunks. However, several encased trees were examined near the Red Zone in March, 1982, and this premise was not substantiated. Trees freed from mud had died, whereas those with a mud coating were still living. As of May, 1982, trees with mud were still healthy green and had new green growth tips. Trees along the Cowlitz and Columbia rivers were salvaged to create spoilage sites. The spoils became levees and dikes that protected properties along the river banks during flooding, especially near precarious river bends.

After the initial dredging and spoilage storage operations on the Columbia River were completed, positive measures were needed to protect against future sediment deposition into the navigation lanes. The May 18 and 19 mudflows demonstrated what effects there could be on these lanes from severe disturbances within the Toutle River drainage basin, which drains an area of 1326 km^2 (512 sq miles) (Pruitt et al. 1980). The headwaters and upper reaches of the North Fork Toutle, particularly north and northwest of the caldera, are primary sediment sources. Therefore, it was and still is necessary for the Corps of Engineers to continue evaluation of the effects of the volcanic damage on the drainage basin and its excessive sediment yield into the Columbia, and to take appropriate measures toward reducing future damage.

To prevent the possibility of another log raft similar to that of May 18 from floating down the Toutle and to prevent massive sediment transfer, the Corps of Engineers recommended construction of the North and South Forks Debris Restraining Structures, designated DRS N-1 and DRS S-1 (Figs. 3.1, 3.13, 3.14, and 3.15) (see n. 2). These structures were expected to restrain almost maximum bed load sediment and permit suspended load passage until sufficient vegetation returned to the blast zone and upper reaches of the North and South Fork Toutle and Green rivers. A 5-yr period was considered reasonable for revegetation of the upper basin, assuming no further destructive volcanic activity occurred. After 5 yr, no significant vegetation had occurred that could retard heavy siltation movement in the Toutle. The Corps also recommended maintenance dredging of DRS N-1 and DRS S-1 reservoirs after each storm flood discharge and repair of any structural breaches (Fig. 3.13–3.16).

FIG. 3.13   Debris Restraining Structure DRS N-1, north side (2/4/81). North spillway of rock gabion sections, 396 m (1300 ft) in length and 10.7 m (35 ft) in height and width each. Note delta created by Hoffstadt Creek (lower right corner). Built for Corps of Engineers. Altitude 1830 m (6000 ft), view looking west–northwest. Location DD on Fig. 3.1.

FIG. 3.14   DRS N-1 north spillway, looking downstream (3/5/82). Breaching of 2/82 completely destroyed north spillway. Note deep head cutting (>12 m [38 ft]) into Hoffstadt Creek channel. Extensive overland flow in reservoir flood plain (right lower corner). Altitude 1370 m (4500 ft), view looking northwest. Location DD on Fig. 3.1.

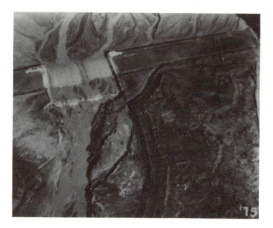

FIG. 3.15   DRS N-1 south spillway, looking upstream to
reservoir (3/5/82). Southside downstream channels have 90°
cliffs. Major eruptive mudflow of 3/19/82 breached the
11.9 m (39 ft) embankment, created 12.2 m (40 ft) wide path
on south side adjacent to hill. See Fig. 3.16. Note construction
road on right side. Altitude 1065 m (3500 ft), view looking
east. Location DD on Fig. 3.1.

FEDERAL MONITORING AND AIR NATIONAL
GUARD RESCUE

Hourly satellite NOA GOES and NASA U-2 aerial photographs were
transmitted to state and county officials as the two major floods of
May 18 and 19 were developing in and beyond the Toutle River system
(O'Lane 1980). These periodic photographs provided accurate data
concerning the rapidly changing events on and around the rivers.
Thus, as mudflows began in the headwaters of the North Fork Toutle
and advanced into the downstream channel, officials were able to warn
Weyerhaeuser's Camp Baker lumber personnel and other persons
near the river banks to move upslope away from the advancing flows.
This unique satellite source, along with National Weather Bureau
information, was part of a preestablished River Flood Advance Warn-
ing System (Cummans 1981). This system was instrumental in saving
many threatened lives at that time.

    Soon after the major eruption of May 18 had subsided and the
blast area was considered accessible, Air National Guard helicopters
flew up the valley on several successful rescue attempts. In retrospect
(according to one pilot), if they had known about the extreme temper-

FIG. 3.16   DRS N-1 breached embankment, south side
(6/6/82). Embankment supporting riprap and crest wall
(steel sheeting) washed out by 3/19/82 flash flood. Bank in
upper portion is the same as that in Fig. 3.15 on right upper
corner. Location DD on Fig. 3.1.

atures above 117°C (350°F) of the avalanche and pyroclastic flows near
the volcano, they would have acted more cautiously.

ASSESSMENT AND EVALUATION OF
DEVASTATED AND AFFECTED AREAS

SCIENTISTS ENTER BLAST ZONE

Several days after eruption, when the blast zone north of Mount
St. Helens was determined to be relatively "safe," geologists entered
and installed instruments for evaluating the damaging alterations to
the mountain and for monitoring future volcanic activity. Seismolo-
gists were able to study, warn others, and predict dangerously high
volcanic activity in the blast area (Miller et al. 1981). Immediately after
the major eruption, the only feasible photographic reconnaissances
were flights by helicopter to record existing, as well as changing,
topography and geomorphology of the devastated region around the
caldera. After the general area cooled to a workable temperature,
many scientists from various organizations set up instruments to study
ground and subsurface temperatures; to measure ground surface
deflections; to determine hydrologic characteristics of rainfall amounts

and intensities, surface water runoff, and erosion; and to record weather data. As the ground grew cooler, biologists investigated and sampled many ponds and lakes, and botanists studied the damage to flora and fauna and its return. Without these early endeavors by scientists, a large store of basic and fundamental knowledge of the early life processes in devastated zones would have been irretrievably lost, along with a comparative basis of their regeneration.

ASSESSMENT OF TOUTLE RIVER DRAINAGE BASIN CHARACTERISTCS

A task force sponsored by the National Forest Service assessed the devastated Toutle River drainage basin, with particular emphasis on the blast zone. This task force consisted of specialists in forestry, geology, and engineering. Using field surveys, tests, and analyses, they identified those characteristics of the basin that would be affected by rainfall and runoff erosion. The single most important characteristic identified was the large denuded area of 391 km^2 (151 sq miles) having steep (>30%) sloped mountainous topography and an annual precipitation of 287 cm (113 in.) (Pruitt et al. 1980). The results of this task force provided the basis for monitoring the blast areas by periodic flight and ground photographic reconnaissance (Table 3.1).

To alleviate concern about the high annual precipitation, the task force determined the following: (1) sediment yield and rate; (2) mountain slope gradients; (3) areas completely denuded, or partially denuded and containing blown down or standing singed trees; (4) position of blown down trees with respect to the slope as either parallel (acting as water shoots) or perpendicular (acting as sediment barriers); (5) drainage morphology and patterns such as rills, gullies, channel and canyon, and dendrites; (6) vegetation, slope material, soil, and aquifer characteristics of permeability and porosity; and (7) precipitation amount and intensity. The task force concluded that the most important topographic and sediment characteristics affected by erosive forces in this drainage system were channel and gully formation, overland flow, and large sediment yield. Stream action drainage patterns rapidly appeared, with severe dissection of the volcanic deposits into dendritic forms on steep slopes (Figs. 3.6 and 3.7) and braided forms on low gradient reaches (Figs. 3.9 and 3.17). On the North Fork Toutle, extensive head cutting developed in deposits of the rock avalanche and pyroclastic flow, beginning at the distal edge, advancing upstream, and ending at the mainstream channels. New channel morphology resulted from bank failures and head cutting by lesser streams. Advanced failures carved out gullies and channels. These failures typically developed in silt, clay, and mud alluvial

FIG. 3.17   Upper North Fork Toutle, 11 km (7 mi) downstream from Maratta Creek (2/4/81). Note "perched" debris deposits impounding unnamed creek on north bank (upper left corner); deposits partially breached by abrasive ash-filled runoff into four linked ponds; flow developing meanders and braided channel drainage patterns. Altitude 1360 m (4500 ft), view looking east–northeast. Location FF on Fig. 3.1.

deposits, indicating a medium to low gradient channel. Braided streams and ponding were prevalent in the hummocky region of the upper reaches of the North Fork Toutle valley (Figs. 3.17 and 3.18).

BLAST ZONE CHARACTERISTICS

North of the mountain, in the denuded blast zone, the steep slopes (>30%) and high elevations of 1341 m (4400 ft) make up approximately 76% of the topography (332 km², or 82,000 acres). The angle of repose is 30% for this soil and rock. Although a snowline over 1341 m (4400 ft) permits vegetation growth, the loss of soil and steep relief of this area prevents a rapid recovery of vegetation and hillslope stability. In 24% of the area (87 km², or 21,500 miles), the slopes can support vegetation because adequate nutrient soils, herbaceous rhizomes, seeds, and less than 15 cm (6 in.) of ash remain. In areas where ash is more than this thickness, it becomes an impervious layer, forcing a large surface runoff into overland flow during intense rainfall.

FIG. 3.18   Ponding of North Fork Toutle, 24 km (15 mi) from caldera (9/9/80). Mile-long formed ponds were controlled when drained by Corps of Engineers (photo courtesy of Corps of Engineers). Altitude 910 m (3000 ft), view looking south. Location EE on Fig. 3.1.

Most leeward slopes that were covered by snow during the eruption are supporting new vegetation. Areas under those fallen logs located perpendicular to the slopes and partially aboveground offered extra protection to existing rhizomes and seeds and are now flourishing. The most successful new growth appears in the center of rills, where runoff has removed the ash and exposed preeruption soils. Vegetation has returned to the lower two-thirds of south-facing slopes on the windward side, where ash was blown away. Close to the caldera, the upper third of south-facing slopes remains badly denuded. Slopes farther away received large amounts of organic debris and pyroclastic and ashfall deposits.

Avalanche and pyroclastic deposits in the upper reaches of the North Fork Toutle formed special "perched" deposits (Fig. 3.17), which impounded many north–south flowing tributaries. Several small depressions among the hummocky deposits contain stagnant pools colored with green, red, or tan bacteria. The Coldwater, South Coldwater, and Castle creeks were impounded and now form lakes and constitute significant water storage areas (Fig. 3.19). Because of the loss of vegetative cover and subsequent water holding capacity in the Toutle drainage basin, water is primarily retained by these reservoirs. These serve a vital role until cover is reestablished.

The topography and geology of the areas surrounding Mount St. Helens are crucial to an understanding of their geomorphology and hydrology. The critical characteristics to be considered are the following: (1) the high relief of the Mount Margaret area (Fig. 3.1, location PP), the ridge east of Spirit Lake, and of Coldwater Ridge west of Spirit Lake (>30% slopes); (2) the high relief of the banks of the Coldwater and Castle lakes (>30% slopes) (Fig 3.1, location GG, and Fig. 3.19); (3) downslope movement of sediment and ash on slopes bared to bedrock surrounding Spirit and Coldwater lakes; (4) the high annual precipitation of 254–305 cm (100–120 in.); (5) the high rainfall intensity of 1.91 cm/hr (0.75 in./hr); (6) the high annual inflow of 216 km²/m (175,000 acre-ft); (7) the high annual sediment delivery of 99 km²/m (80,165 acre-ft); (8) the total capacity of impoundments of 780 km²/m (633,000 acre-ft); and (9) the poorly consolidated soils with low or nonplastic cohesion on steep slopes. The river channels are subject to rapid and deep eroding from rill, gully, and canyon erosion (Table 3.2) (Pruitt et al. 1980).

GREEN RIVER DRAINAGE CHARACTERISTICS

The Green River drainage basin was unaffected by the major eruption in its lower reaches, but it was devastated in its upper reaches, especially near the headwaters of Grizzly Creek. The major blast cloud swept trees down in a 4.0-km (2.5-mile) wide path beyond Ryan Lake. The wave continued north about 3 km (2 miles) down the headwaters of Quartz Creek. A halo of singed trees 0.4–1.2 km (0.25–0.75 mile) wide formed along the edge of the blown down trees. The standing dead timber represented 30% of the total damage. Several other oddly formed blowdown and seared forest patterns occurred in this drainage, apparent from the irregular boundaries of the blast and halo zones identified in Figure 3.1.

The thickness of the tephra deposits varied from 8.9 to 10.2 cm (3.5 to 4 in.). Mudflow deposits thought to have originated from rapid snowmelt during the major eruption moved down the upper Green River, Grizzly Creek, and Miners Creek, forming alluvial fans in the mainstream of the Green. In the period since the eruption, 1.5–3 m (5–10 ft) depth of tephra sediment has been deposited in the pool habitat of the upper Green River. Within the blast area, blown down trees spanned stream channels in dangerous densities, damming valley streams with low gradients. When winter and spring storm runoff occurred, the downed trees and other woody debris were transported downstream in the peak flows. This debris was trapped by and contributed to a dam forming a lake near Quartz Creek (Fig 3.1, location QQ). Although there were many downed trees on slopes of >30% gradient, sediment yield and its transport were lower here than in other geologically comparable basins because the fallen trees held back

FIG. 3.19   Coldwater and South Coldwater lakes (3/5/82). Coldwater Lake ponding behind pyroclastic debris impoundment dam; no lake forming in south Coldwater Creek; note multiple ponds and phreatic craters. Altitude 2270 m (7500 ft), view looking north. Location GG on Fig. 3.1.

the sediment. Nevertheless, high sediment yield occurred from the many previous clearcuts in the basin because little or no woody debris existed on their surfaces.

Although the upper basin was badly damaged, no debris from the pyroclastic flow entered and clogged the channels. Water quality may improve here sooner than in the upper Toutle (Klein 1981). In badly damaged areas, however, the downed trees and tephra of ashfall greater than 15.2 cm (6 in.) thick will retard quick recovery. The lower reaches of the Green River were not affected by the eruption, and this area remains as a living, active forest greenbelt restraining sediment movement. In these forested streams, the lower temperatures cooled the open and heated water from the hot denuded upper reaches. The western Green River drainage basin lies adjacent to the North Fork Toutle, and only the devastated upper Green, near Quartz Creek, requires cautious use. If logging must continue, it should be done only on riparian and adjacent slopes, no farther than 61 m (200 ft) from the river. After logging, these areas will require fertilization and planting with seedlings. Debris fences, consisting of logs staked perpendicular to the slopes, should be built on preeruptive clearcuts to restrict severe downslope erosion (Pruitt et al. 1980). When filled, the debris-fenced area should be seeded and fertilized to promote bank stabilization.

ASSESSMENT AND EVALUATION OF DEVASTATED AND AFFECTED AREAS 21 MONTHS AFTER THE ERUPTION

CONDITIONS ALONG THE NORTH FORK TOUTLE

On March 5, 1982, turbid waters still moved swiftly into the Tower Bridge reach of the Toutle River. From the airplane, the afternoon sun glinted from standing wave surfaces. Large boulders, gravel, and pebbles were now a stationary part of the river bed. Moving upstream, past the confluence of the two forks, channels were changing from a sinusoidal pattern in narrow and medium gradient passages to meandering braided streamlets and channels in broad, nearly level plains. Midstream in these reaches, there were many sand and gravel bars. Lining the river banks were right-angled cliff edges, created by blocks of flood plain sediment breaking off and falling into moving waters.

Closer to Mount St. Helens, the waters of the Toutle River appeared more turbid. At the headwaters of the North Fork Toutle, the surface runoff, which was filled with suspended ash particles, silt, and mud, continued from the previous day's heavy, intense rainfall. This suspended load was formed from poorly consolidated layers of erosive flood plain sediments and the eroded rill and gully material from the steep slopes surrounding the blast area.

At Coal Bank (Fig. 3.1, location I), a replacement bridge was under construction. The abutments were reinforced along the base of the cliff with 183 m (600 ft) of short pilings in an attempt to stabilize the bank from future slumping. About 27.5 m (90 ft) of bank material had fallen as colluvium and alluvium in large blocks, feeding the high-velocity flow of the storm's bankful discharge moving down the Toutle. Farther upstream, previously braided stream channels had become sinuous, leaving thin narrow streamlets flowing near the banks, again cutting and filling the flood plains. More than 30.5 m (100 ft) of the flood plain deposited on May 18, 1980, had eroded for 0.8 km (0.5 mile) above the Coal Bank bridge (Fig. 3.1, location AA) and was transported into the Cowlitz and Columbia rivers during the first year after the eruptions by three intense rainstorms. Bank failures, incised sidestreams (Fig. 3.20), and subsequent head cutting of these loosely

FIG. 3.20   Incised streamlet in 5/18/80 mudflow flood
plain (6/6/81). Note two distinct layers in 10,000-yr event
flood plain alluvium: poorly sorted, dark-colored, volcanic
pyroclastic (andesite and dacite) deposits on bottom, overlain
with fine-grained, well-sorted silt, ash, and mud, indicating
two different flood deposits. Rounded pebbles of pumice and
pumiceous rocks on either side of creek. View looking north.
Location AA on Fig. 3.1.

compacted and many graded debris strata of dacite, andesite, basalt,
and ash provided much material to the storm runoff discharge filling
the North Fork Toutle from bank to bank. Fed by surface runoff
during each storm, these sidestreams sliced channels deeply into flood
plain deposits, increasing mainstream channel siltation and suspended
load. In coming years, huge meandering streams will probably dominate the geomorphology of this reach of the North Fork Toutle until
vegetation begins to restrain erosion.

     Near the High Bridge (Fig. 3.1, location BB, and Figs. 3.21–
3.24), the extensive salvaging of mud-encrusted riparian trees appeared to exceed reasonable limits. This high bank and narrow gorge
of the North Fork Toutle region took the brunt of the May 18 high
flood, and many bankside trees were encrusted with a mud layer
(Fig. 3.22). Most trees were located on the northern side of an ancient

flood plain. Unfortunately, logging was not stopped after removal of these trees but was continued up the steep slopes, until a total of 0.16 km² (46 acres) was cleared (Figs. 3.23 and 3.24). According to standard logging practice, clearing this many acres is considered a practical minimum. Nonetheless, from the viewpoint of desired water retention in this critically damaged riparian zone, only those trees absolutely necessary (mud-encrusted ones) should have been removed. A greenbelt should have been left along each creek, stream, and river in the upper reaches of the basin. The steep slopes in this gorge permit rapid slumping of water-soaked soil and sediment. The trees on these slopes, had they remained, would have dispersed raindrops during each rainfall, increased infiltration into the soil, and prevented soil compaction, slumping, or mass erosion. The high surface runoff after each storm would have been dissipated into subsurface flows. Perhaps recognizing their error, the responsible lumber company's personnel immediately replanted after clearing, although 5 yr are permitted by Forest Service regulations for replanting a clearcut. In June 1982, however, the adjacent downstream section on the north side of the river was clearcut, permitting new rills and gullies to form on these slopes (Fig. 3.1, location BB).

Several logging companies were clearing blown down trees up-steam. The prime company salvaged about 60–70% of the singed and blown down trees from their land located in the northwestern portion of the blast area. They did leave a token 0.16 km² (40 acre) plot of singed trees just upstream from the Hoffstadt Creek confluence with Bear Creek (Fig. 3.1, location N).

TREE SALVAGING IN THE TOUTLE DRAINAGE

The following questions were posed by several congressmen at a public hearing held in Vancouver, Washington, on April 4, 1982:[3]

1. Since two-thirds of the damaged trees in the northwestern blast area have been removed, should the remainder also be removed?

2. Should the 200–300-year-old trees from the Valley of the Giants in the Green River area be removed?

3. Should the Mount St. Helens devastated area be set aside as a National Park, National Monument, or Volcanic Research Area, or should it revert to a preeruption status?

The majority of scientists present at the hearing stated that both the ancient trees and more than a token acreage of destroyed trees should be preserved. County and logging industry officials countered

FIG. 3.21   North Fork Toutle area north of Kid Valley
(9/2/80). Weyerhaeuser's Camp 19 at center; High Bridge
SR 504 at upper right. Note protective greenbelt on west
bank of river. Altitude 2120 m (7000 ft), view looking south.
Location BB on Fig. 3.1.

FIG. 3.22   High Bridge SR 504 (9/28/80). Second growth
Douglas Fir greenbelt shown; note high flood stage mark line
on trunks. Closeup of greenbelt area in Fig. 3.21. View
looking west. Location BB on Fig. 3.1.

FIG. 3.23   High Bridge SR 504 (3/28/81). Note clear-cutting of greenbelt, instead of removal of mud-encased trees only. Compare with Fig. 3.22. View looking west. Location BB on Fig. 3.1.

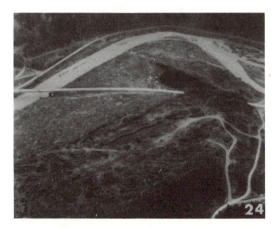

FIG. 3.24   High Bridge SR 504 (3/5/82). Aerial view of area around Fig. 3.23; extent of clear-cut is shown. With slope gradient>30%, gully erosion and mass movement of sediment are occurring. Altitude 610 m (2000 ft), view looking south. Location BB on Fig. 3.1.

that all of the trees should be salvaged and only a limited acreage of 340 km² (84,700 acres) should be set aside for research and a future preserve. It is interesting that the average nonaffiliated citizen agreed with the scientists. They preferred that a large number of acres (>800 km², or >200,000 acres) should be reserved for research and become a National Monument after the Volcanic Research Area status had expired. The Research Area status would remain for 5 yr after any scientifically identified volcanic activity.

TREE SALVAGE RECOMMENDATIONS

To minimize and control lower basin siltation, as much sediment as possible should be held *in situ* in the upper reaches of the Toutle River drainage basin. Restraining the debris could reestablish stability and reduce the heavy runoff erosion after each rainstorm. Since the denuded slopes immediately north of the caldera will probably yield large amounts of sediment in the next several years, a concentrated effort should be made to limit or control large sediment yield from these surrounding areas. Blown down trees act as debris fences, reducing downslope movement of volcanic debris and ashfall material from these denuded hills. In this region, gullies have incised erosional paths through these layers and exposed preeruptive native nutrient soil. This encourages new vegetation. Nevertheless, the impervious layer formed by the ashfall of more than 15 cm (6 in.) near the mountain (Sarna-Wojcicki et al. 1981) remains a serious problem. As a partial solution, the fallen trees, limbs, and debris should be left to disperse raindrops from each storm and to slow down runoff. The slower moving water will erode the ash layers gradually and will eventually infiltrate into subsurface flows and aquifers. The water will also develop small rills and gullies through the ash, exposing more native soil and promoting vigorous revegetation. Once these soil ribbons are opened to the elements, seeds, rhizomes, and vegetation will take hold (Sedell et al. 1980). Animals and birds will return with the advent of vegetation and insects and add fertilizer that will enrich the soil lacking in phosphorus and nitrogen. Other animals will help mix the ash into the soil, thereby introducing new minerals. With the continued presence of animals, vegetative stability will return to these extremely poor soils. As the partially stripped blown down trees weather and decay, their organic matter will also add mulch and nutrients, further enriching the soil.

At various public meetings held concerning the status of the blast area and destroyed tree zone, a shared belief became evident. Many people believed that, if the destroyed blown down and standing trees remained, insects and diseases would penetrate them. These trees, in turn, would endanger the adjoining living ones as well. On the other hand, scientists suggested that insects and animals attacking these trees would serve as food for larger animals and birds in the normal food chain. In addition, several people thought that these trees would provide fuel for a catastrophic forest fire. The majority of singed trees, however, have been removed from the blast area, thus diminishing the fuel needed for either small forest fires or a catastrophic forest fire. Actually, this threat of fires was purely hypothetical because precipitation is high in the blast zone, which tends to maintain fuel moisture and reduce fire danger.

For the above reasons, preserving singed and blown down trees became even more significant because only one-third of the destroyed trees remained. Retaining these trees in the blast zone becomes essential for comparisons with healthy trees in adjacent zones while the devastated forest regenerates. Therefore, my recommendation was that more than 1214 km² (300,000 acres) should be designated as a Mount St. Helens Volcanic Research Area. This acreage should include not only the blast zone but also the greenbelt corridors on the flood plains of all rivers draining Mount St. Helens.

On August 26, 1982, President Reagan signed legislation establishing a 445-km² (110,000-acre) Mount St. Helens National Volcanic Monument. This acreage does not include any downstream river flood plains formed by volcanic mudflows.

ASSESSMENT OF SEDIMENT
AND WATER RETENTION
NEAR THE VOLCANO

The following are abbreviated progress reports of natural and human activities in some important areas of the North Fork Toutle for the time period from 1980 to 1982. These evolving natural processes and their geomorphological changes are described in Table 3.1.

SPIRIT LAKE

There was a significant increase in the stored water volume. The southeast lobe was greatly enlarged in March, 1982, when ash, pyroclastic deposits, and mud created a dam (Figs. 3.6–3.8).

STORAGE LAKES OF COLDWATER
AND CASTLE CREEKS

Many rills and gullies developed dendritic drainage patterns through-out the debris and ash layers. On the west-facing hillsides, regeneration of a minimal amount of plant growth began to appear as material collected behind debris fences created by blown down trees. The surrounding hillsides remained severely denuded, limiting plant rejuvenation. The initial pyroclastic debris impoundment remains structurally sound, permitting runoff waters to raise lake volumes to levels established by the Corps of Engineers and scientists. By June, 1982, drainage channels through bedrock from these lakes to the North Fork Toutle had been constructed by the Corps to control the lake volumes (Fig. 3.18).

VOLCANIC RAMP NORTH OF CALDERA

An estimated 30-m (100-ft) deep canyon caused by erosive head cut-ting developed in the loosely consolidated pyroclastic debris to the edge of the ramp. It was filled layer by layer with pyroclastic and ash flows from each subsequent volcanic eruption cycle (Figs. 3.3 and 3.4).

NORTH FORK TOUTLE DEBRIS
RESTRAINING STRUCTURE (DRS N-1)
AND RESERVOIR

Control was lost when the south spillway was breached by a mudflow in December 1980. The spillway was replaced by a roller-compacted concrete spillway (Fig. 3.15). The north span rock gabion spillway then failed in the March 1982 mudflow (Fig. 3.14). The southern sidewall near the river bank and the main midspan structural rock embank-ment and crest wall also were breached at this same time (sidewall intact, Fig. 3.15; sidewall breached, Fig. 3.16). Vital human responses that were required included (1) dredging the reservoir basin after sediment inflow of each storm surface runoff, and (2) repairing the breaches to the main structure and spillway after each occurrence. By May 1982, the reservoir capacity had been reduced to 10% because three breached sectors were unrepaired. Funding and activities ceased in early 1982. In 1985, this structure remained unrepaired, but the Corps of Engineers recommended the construction of a new retaining dam below Spirit Lake.

EFFECTIVENESS OF THE DEBRIS
RESTRAINING STRUCTURE (DRS N-1)

In March 1982, no water was ponding behind the restraining structure
(Fig. 3.14) and the reservoir required dredging, whereas in February
1981 (Fig. 3.13), ponding did occur, as it had in November 1980 when
the basin was first completed. Dredging and repair was done for 1.5 yr
until funding stopped. The reservoir slowed downstream bed load
sediment transport.

During the March 19, 1982, eruption, a 12.2 m (40 ft) wall of ash
and mud moved over the volcanic ramp and diminished to a 2.4 m
(8 ft) flash flood overtopping a portion of the DRS N-1.[4] The DRS N-1
served its design objective by reducing the flood wave before it continued downstream. The flood breached the midspan embankment of
the southern structure and the northern spillway. A total of
760,000 m³ (1 million yd³) of debris flowed through the Toutle and
deposited into the Cowlitz River. If this had been combined with the
entrapped amount, the downstream areas would have suffered severe
consequences. Theoretically, the DRS N-1 structure had the storage
capacity to absorb such flash floods without overtopping for at least a
5-yr period if the reservoir was kept dredged.

The next volcanically induced mudflow will probably cause the
existing breach of the DRS N-1 to enlarge and permit embankment
failure to its base on the downstream side. Massive amounts of large-
sized particles and other sediment from the reservoir will be transported into the Toutle, Cowlitz, and Columbia rivers. In addition,
there is a potentially large volume of easily transported sediment in the
upper drainage basin. This mudflow could be two or more times larger
than the initial May 18, 1980, mudflow. To keep this sediment and
debris from discharging into the more populated reaches of the Toutle
River drainage basin, the DRS N-1 should remain operative. Several
additional water retention or settling basins also may be required.
Satellite monitoring keeps the Corps of Engineers informed of the
upper basin status. This is essential to an understanding of the basin's
changing morphology as it affects the DRS N-1 and to an understanding of problems in the lower river basin.

FUTURE ACTIONS

When the DRS N-1 structure was constructed, people living downstream of the volcano gained a sense of security concerning the threat
from massive flooding of the weakly consolidated, large particle debris

during future rainstorms and volcanic mudflows. Keeping the reservoir of DRS N-1 cleared, however, became an unacceptable financial burden, and dredging and repairs were halted. A search for a better long-term solution was required for the blast zone itself. An intensive research study was initiated to determine the hydrologic and geomorphic factors affecting the annual sediment yield, movement, and slope stability of the upper drainage basin.

Natural storage reservoirs were created when the avalanche and the debris flow impounded the waters of Coldwater and Castle creeks (Table 3.2). The hills enclosing these reservoirs are precariously steep and are denuded of soil and vegetation. They are capped with a thin, impermeable veneer of ash (>15 cm, or 6 in.). This region is also subject to high annual precipitation and intense rainstorms. The debris-dammed lakes themselves presented other fundamental problems. The structural integrity, permeability, and endurance of the debris dams were extremely questionable (Jennings et al. 1981). Nevertheless, by May 1982, these natural dams were still structurally intact. A constant threat of failure existed by piping or overtopping of the dams and flash flooding down the North Fork Toutle.

These two problems, piping and overtopping, are dependent on the condition of the preeruptive buried sand and gravel beds, storm runoff inflow rate, and maximum capacity of each lake. If these beds have high permeability, are sound, and have sufficient depth to permit adequate percolation of inflowing surface runoff, then a satisfactory lake storage level will be maintained. If pressure increases as a result of rising waters, and if these beds permit excessive seepage through a weak layer or along an impermeable layer, the result would be piping through the debris dam, which could cause a catastrophic failure. If the measured flow rates increased in conjunction with filling beyond the capacity of the lakes, overtopping could cause excessive catastrophic erosion and failure of the dams. Personnel from the Forest Service Task Force, U.S. Geological Survey, and U.S. Army Corps of Engineers performed extensive field studies on these natural storage lakes and determined the probability of piping or other failures (Jennings et al. 1981). As of mid-1982, these impounded structures appeared to be structurally sound, although their continued durability remains questionable.

The overtopping of both Castle and Coldwater lakes are controlled with the construction of drainage channels through bedrock from the lakes to the North Fork Toutle. These channels maintain a "safe" water level, barring any mammoth storm or violent volcanic activity that would cause rapid filling beyond the capacities of the lakes.

In August 1982 very rapid erosion of the volcanic debris impounding Spirit Lake caused the Corps of Engineers to issue a statement of emergency. Emergency funds became available to allow both the Corps and the Federal Emergency Management Agency (FEMA) to begin pumping operations to prevent the lake from rising beyond safe levels and the destruction of the debris dam impounding these waters. Pumping operations were performed for about 2.5 yr on a 24-hr, 7-day schedule. In April 1985 the Corps dedicated a tunnel to drain water from Spirit Lake. It was bored through bedrock for 2578 m (8460 ft) from the southwest side of Spirit Lake to the South Coldwater Creek valley. Draining water from Spirit Lake would then continue through the Coldwater Lake channel into the North Fork Toutle.

The following additional hydrologic characteristics should be monitored in the blast zone: (1) infiltration capacity through the sediments; (2) sediment movement and yield based on long-term measurements to determine slope stability, because measurements made in mid-1982 were on a short-term basis during a single month; (3) drainage pattern development; (4) sediment yield from runoff rill and gully erosion; (5) sediment yield from overland erosion; (6) usefulness of levees, dikes, and storage basins, especially in the lower reaches of North and South Fork Toutle and the main reaches of the Toutle, Cowlitz, and Columbia rivers; and (7) consequences of continued filling of the Toutle and Cowlitz river channels which reduce channel capacities.

Knowledge of these storage lake parameters and drainage basin characteristics are essential and require updated reevaluation so that researchers can accurately observe, deduce, and report the latest geomorphic and hydrologic cycle in this basin for accurate predictions of future changes, especially the detrimental ones. The possibilities of future eruptions, debris flows, and mudflows are well documented (Crandell et al. 1975, Crandell and Mullineaux 1978, Korosec et al. 1980) and must be carefully considered in future plans.

RECOMMENDATIONS

An engineering geology recommendation for the entire blast zone is as follows. Every effort should be made to hold sediment *in situ* in the blast zone and on the river flood plains. This effort should continue until the region becomes stabilized through revegetation. To restrain the existing mammoth amounts of sediment in place, one or more of the following measures might be tried: (1) build debris fences on steep slopes with onsite logs or those retrieved from the river bed; (2) seed with indigenous grasses, fertilize, and plant with tree seedlings that are advanced enough to withstand animal browsing; (3) halt all clear-cutting and tree salvaging; (4) using rock from nearby avalanche and debris flow blocks, build rock barriers across all creeks as debris retention sites, especially across the channel of the North Fork Toutle, at regularly spaced intervals (to be determined on the basis of the sediment yield from the microdrainage area of each barrier); and (5) create settling or holding ponds behind each of the above barriers. As these barriers fill and become less porous, they will filter the suspended load and restrain large particles from the bed load. Even if these recommendations are effective for only a few years, it may give time for stability to be established in the Toutle River drainage basin. This, of course, assumes that Mount St. Helens will not produce another catastrophic eruption in the near future.

CONCLUSIONS

Monitoring the response of the Toutle River system to environmental and human forces using aerial and ground photographic surveillance permits any researcher to study important geomorphic and hydrologic changes. Certain changes could have significant long-term consequences. As seen from a flight on May 23, 1982, the devastating effects of the March 19, 1982, volcanic mudflow were vividly apparent. The U.S. Army Corps of Engineers restraining structure had its storage capacity exceeded, and the spillways were breached at three locations. Unless these spillway breaches are repaired and the reservoir storage capacity reestablished, the threat of large sediment yield from the upper drainage basin and accumulated debris and sediment in the DRS N-1 reservoir will become a reality with each bankfull discharge. Funds to restore the storage capacity and to repair the structure should be reinstated; if this is not done, measures of equal worth should be instituted. Money spent at this location now would amount to much less than might be needed in the future for downstream dredgings of the Cowlitz and Columbia rivers.

After each intense rainfall and runoff cycle, similar channel bed load deposits from eroding upper basin flood plains are filling the lower reaches of the Toutle and Cowlitz. Two years after the major eruption, an estimated 340,000 m³ (440,000 yd³) of sediment has been eroded and transported downstream from a 1-km (0.6-mile) long flood plain of the North Fork Toutle. Funds are needed for periodic dredging of these bed load deposits to prevent damage to the Columbia River navigation lanes and loss of water supplies, sewage, and storm drain systems for the towns along the Toutle and Cowlitz rivers.

Tree planting and frequent fertilization should continue in the upper drainage basin after salvage operations are completed, especially in the blown down and singed tree zones. This action will help reestablish the basin's vegetation and slope stability, as well as reduce sediment yield to nearby streams and rivers. The successful efforts performed by grass seeding and fertilizing to reduce silting in the upper basin should be continued. Movement of large sediment particles and flash flooding has been prevented by controlled draining of volcanic ponds and impounded lakes near the volcano. Dredge spoils creating levees, dikes, and basins in the lower Toutle have minimized danger to human life along the river banks and have maintained channel capacities through the Toutle, Cowlitz, and Columbia rivers. Another successful endeavor has been grass seeding of large spoilage mounds located along the banks of the Cowlitz River, a welcome sight of green instead of the previous ash-gray mounds.

Nature is doing its part. Many alder seedlings have germinated and are becoming sizable on the flood plains and banks of rivers flowing from the volcano. Many other tree seedlings, bushes, and other vegetation are returning to the entire basin. Active erosion in the blast zone removes excess ash and exposes soil for plant growth. Because of the above average annual precipitation and mild winters, revegetation has encouraged the rapid return of animals and birds. Nature's healing processes are in force, creating a new environment around Mount St. Helens.

ACKNOWLEDGMENTS

Many thanks are given to my husband Phil and son Joseph (my field trip assistant). Special thanks go to E. W. Brogren, aeroengineer and self-taught geologist, who made a Cessna 172 plane always available. He also participated in discussions of the pyroclastic flow dynamics and the entire effort, and reviewed the original manuscript. Other reviewers were Frank Lewis and my daughter Pamela, both geologists, and Curtlan Betchley, geophysicist. Their collective criticisms helped to solidify and condense the paper. Thanks are also extended to Dr. David E. Bilderback who initiated the request, encouraged me, and edited this paper. Many thanks are extended to Mrs. Shirley Boond, who typed the manuscript, and Mrs. Joy Samora, who helped with the artwork. Appreciation is extended to Robert E. Norris, geologist, University of Washington Geophysics Volcanic Watch, for continued notification of significant volcanic activities and as field associate with Ed Rozmyn. Without the assistance of the Public Relations Office of the U.S. Army Corps of Engineers, especially Jim Addison, and Lee Fairchild of the University of Washington, much useful data would not have been obtained.

NOTES

1. Avalanche–debris flow dimensions were measured from the map, Mount St. Helens and vicinity, Washington–Oregon, 45121-H8-TM-100, Gifford Pinchot National Forest, Pacific NW Region, USDA Forest Service, U.S. Geological Survey and Department of Natural Resources, State of Washington, March 1981.

2. Data pertaining to efforts of U.S. Army Corps of Engineers, Portland District, Oregon, are taken from the following Fact Sheets: March 1981, May 1981, November 1982, and April 1985.

3. Meeting held at Vancouver, Washington, on April 4, 1982: "Mount St. Helens Field Hearings." References: Bills-HR: Washington State Governor—106,000 acres, 5773; U.S. Department of Agriculture, Gifford Pinchot National Forest Service—84,700 acres, 5787; and Representative Don Bonker—216,000 acres.

4. Debris Restraining Structure data from Fact Sheets, U.S. Army Corps of Engineers, Portland District: (1) Mount St. Helens Recovery Work Cowlitz and Toutle Rivers, March 1981; (2) Mount St. Helens Restoration, May 1981.

LITERATURE CITED

BINGHAM, R. 1980. Explorers of the earth within. *Science 80* 1 (6):44–55.
CRANDELL, D. R., and D. R. MULLINEAUX. 1978. Potential hazards from future eruptions of Mount St. Helens volcano, Washington. *U.S. Geol. Surv. Bull. 1383-C*, 26 pp.
CRANDELL, D. R., D. R. MULLINEAUX, and M. RUBIN. 1975. Mount St. Helens volcano: recent and future behavior. *Science (London)* 187:438–441.
CUMMANS, J. 1981. Mudflow resulting from the May 19, 1980, eruption of Mount St. Helens, Washington. *U.S. Geol. Surv. Circ. 850-B*, 16 pp.
DECKER, R., and B. DECKER. 1981. The eruptions of Mount St. Helens. *Sci. Am.* II:68–80.
HOBLITT, R. P., D. R. CRANDELL, and D. R. MULLINEAUX. 1980. Mount St. Helens eruptive behavior during the past 1500 years. *Geology* 8:553–559.

JANDA, R. J., K. M. SCOTT, M. NOLAN, and H. A. MARTINSON. 1981. Lahar movement, effects and deposits. In *The 1980 Eruptions of Mt. St. Helens, Washington*. Lipman, P. W., and D. R. Mullineaux (eds.), U.S. Geol. Surv. Prof. Paper 1250, pp. 461–478.

JENNINGS, M. E., V. R. SCHNEIDER, and P. E. SMITH. 1981. Computer assessments of potential flood hazards from breaching of two debris dams, Toutle River and Cowlitz River systems. In *The 1980 Eruptions of Mt. St. Helens, Washington*. Lipman, P. W., and D. R. Mullineaux (eds.), U.S. Geol. Surv. Prof. Paper 1250, pp. 829–836.

KERR, R. A. 1981. Mount St. Helens. *Science (London)* 212:1258–1259.

KIEFFER, S. W. 1981. Fluid dynamics of the May 18 blast at Mount St. Helens. In *The 1980 Eruptions of Mt. St. Helens, Washington*. Lipman, P. W., and D. R. Mullineaux (eds.), U.S. Geol. Surv. Prof. Paper 1250, pp. 379–400.

KLEIN, J. M. 1981. Some effects of the May 18 eruption of Mount St. Helens on river-water quality. *In The 1980 Eruptions of Mt. St. Helens, Washington*. Lipman, P. W., and D. R. Mullineaux (eds.), U.S. Geol. Surv. Prof. Paper 1250, pp. 719–731.

KOROSEC, M. A. 1981. Mount St. Helens fact sheet, Mount St. Helens dome status, and river cleanup continues. *WN Geol. Newsletter, WN State DNR* 9(2):15–18.

KOROSEC, M. A., J. G. RIGBY, and K. L. STOFFELL. 1980. *The Eruption of Mount St. Helens, Washington, Part 1: March 20–May 19, 1980*. Inf. Cir. 71, WN State DNR, 26 pp.

LOMBARD, R. E., M. B. MILES, L. M. NELSON, D. L. KRESH, and P. J. CARPENTER. 1981. The impact of mudflows of May 18 on the lower Toutle and Cowlitz rivers. In *The 1980 Eruptions of Mt. St. Helens, Washington*. Lipman, P. W., and D. R. Mullineaux (eds.), U.S. Geol. Surv. Prof. Paper 1250, pp. 693–699.

LOOMIS, B. F. 1926. *Eruptions of Lassen Peak*. Loomis Museum Assoc., Lassen Volcano National Park, California, 100 pp.

McLUCAS, G. 1980. Petrology of current Mount St. Helens tephra. *WN Geol. Newsletter, WN State DNR* 8(3):7–16.

MILLER, C. D., D. R. MULLINEAUX, and D. R. CRANDELL. 1981. Hazards assessments at Mount St. Helens. In *The 1980 Eruptions of Mt. St. Helens, Washington*. Lipman, P. W., and D. R. Mullineaux (eds.), U.S. Geol. Surv. Prof. Paper 1250, pp. 789–802.

MOORE, J. G., and W. C. ALBEE. 1981. Topographic and structural changes, March–July 1980: photogrammetric data. In *The 1980 Eruptions of Mt. St. Helens, Washington*. Lipman, P. W., and D. R. Mullineaux (eds.), U.S. Geol. Surv. Prof. Paper 1250, pp. 123–134.

MOORE, J. G., and T. W. SISSON. 1981. Deposits and effects of the May 18 pyroclastic surge. In *The 1980 Eruptions of Mt. St. Helens, Washington*. Lipman, P. W., and D. R. Mullineaux (eds.), U.S. Geol. Surv. Prof. Paper 1250, pp. 421–438.

MULLINEAUX, D. R., and D. R. CRANDELL. 1981. The eruptive history of Mount St. Helens. In *The 1980 Eruptions of Mt. St. Helens, Washington*. Lipman, P. W., and D. R. Mullineaux (eds.), U.S. Geol. Surv. Prof. Paper 1250, pp. 2–15.

O'LANE, R. G. 1980. Volcanic eruption disrupts air traffic. *Aviation Week and Space Tech.* 1:18–21.

PRUITT, J., J. EDGREN, B. HAMNER, G. HAUGEN, S. HOWES, P. PATTERSON, J. STEWARD, and J. SWANK. 1980. *Mount St. Helens Emergency Watershed Rehabilitation Report*, USDA Forest Service, Pacific NW Region, Portland, Oregon, 96 pp.

RAU, W. W. 1980. Washington coastal geology between the Hoh and Quillayute rivers. *WN DNR Bull.* 72:2–57.

ROSENFELD, C. L. 1980. Observations on the Mount St. Helens eruption. *Amer. Scient.* 68:494–509.

SARNA-WOJCICKI, A. M., S. SHIPLEY, R. B. WAITT, JR., D. DZURISIN, and S. H. WOOD. 1981. Aerial distribution, thickness, mass, volume and grain size of airfall ash from the six major eruptions of 1980. In *The 1980 Eruptions of Mt. St. Helens, Washington*. Lipman, P. W., and D. R. Mullineaux (eds.), U.S. Geol. Surv. Prof. Paper 1250, pp. 577–600.

SCHUSTER, R. L. 1981. Effects of the Eruptions on civil works and operations in the Pacific Northwest. In *The 1980 Eruptions of Mt. St. Helens, Washington*. Lipman, P. W., and D. R. Mullineaux (eds.), U.S. Geol. Surv. Prof. Paper 1250, pp. 701–718.

SEDELL, J. R., J. F. FRANKLIN, and F. J. SWANSON. 1980. Out of the ash. *Am. For.* 86(10): 26–68.

WATT, R. B., JR. 1981. Devastating pyroclastic density flow and attendant air fall of May 18: stratigraphy and sedimentology of deposits. In *The 1980 Eruptions of Mt. St. Helens, Washington.* Lipman, P. W., and D. R. Mullineaux (eds.), U.S. Geol. Surv. Prof. Paper 1250, pp. 439–458.

WINNER, W. E., and T. J. CASADEVALL. 1981. Fir leaves as thermometers during the May 18 eruption. In *The 1980 Eruptions of Mt. St. Helens, Washington.* Lipman, P. W., and D. R. Mullineaux (eds.), U.S. Geol. Surv. Prof. Paper 1250, pp. 315–320.

WORTHINGTON, R. E. 1981. *Mount St. Helens Land Management Plan.* Draft Environmental Impact Statement, USDA, Gifford Pinchot National Forest, Vancouver, Washington, 162 pp.

TABLE 3.1

GEOMORPHOLOGIC–EROSIONAL SITES DESCRIBED FROM
AERIAL AND GROUND PHOTOGRAPHIC RECONNAISSANCES[a]

Site	Processes and geomorphologic changes
Volcanic Ramp (low to high percent gradient slope)	Deep, straight, and sinuous gullies, channels, and canyons. Extensively incised canyons in pyroclastic deposits, intense headcutting toward caldera by channels and gullies.
North Fork Toutle, Upper Reaches (down valley to distal edge of debris deposit; low gradient slope)	Heavily braided stream channels, developing from several rills and gullies forming in silt and clay fill within the hummocky pyroclastic debris layers. Incised channels, 15–46 m (50–150 ft), becoming mainstream bed (near Elk Rock) caused by headcutting upflow. Many phreatic eruptive craters and depressions available for water storage; poorly consolidated hummocky deposits, high infiltration rate.
South Fork Toutle, Upper and Lower Reaches; Lower North Fork Toutle	Sinuous to braided channels developing, assumed on the basis of low-gradient slope. Many sand and gravel bars forming in midstream channels with braided streamlets across bars. Heavily incised rills, gullies, and channels headcutting toward the banks on the flood plains. Large volumes of flood plain sediment transported downstream with each peak flow.
Spirit Lake; Surrounding Hills and Basin (high gradient slope; water level of lake and structural integrity of dam critical; see Table 3.2 for data)	Many rills, gullies, and channels forming on steep slopes, particularly through debris and ashfall deposits, forming alluvial fans. Dendritic drainage patterns continuing to form and vary as long as sediment remains available for mass movement downslope. Movement greatest during intense and heavy precipitation as runoff erosion. Blasted slopes covered with tephra deposits, continuing to move downslope until all loose ash is moved.

MONITORING NEEDED |

TABLE 3.1

GEOMORPHOLOGIC–EROSIONAL SITES DESCRIBED FROM
AERIAL AND GROUND PHOTOGRAPHIC RECONNAISSANCES[a] (continued)

Site	Processes and geomorphologic changes
Coldwater and South Coldwater Basins (extreme gradient slope; water level critical to retain structural integrity of dam; see Table 3.2)	Steep slopes and intense high rainfalls; many rills, gullies, and channels developing; severe headcutting by silt, clay, and mud block collapsing into the drainage. Gully and channel erosion processes producing very high sediment yields. Hillslope material forming alluvial and colluvial fans, removed by stream action in the South Coldwater basin or filling lakes in the Coldwater basin. The severity of these processes is assumed on the basis of volcanic material amount and type on the slopes. Hummocky material and low density deposits of impounded dam subject to collapse and liquification during earthquakes. Further risk of a failure by piping of groundwater through the dam; also gully and channel erosion incisions toward the dam.

MONITORING NEEDED |
| *South Fork Castle* (high gradient slope; water level critical to retain dam integrity; see Table 3.2) | Drainage pattern of rills, gullies, and channels on hillslopes, tempered by more native soils remaining on slopes. Where blast affected slopes, drainage identical to Spirit Lake. Piping of groundwater and overall storage of inflowing water is a function of buried alluvial gravels present prior to blast. (This problem is pertinent to both Coldwater Lake and Spirit Lake and may cause premature dam failure.) If original gravels are buried intact and not filled, the aquifers may exist beneath these reservoirs (porosity and permeability of pyroclastic flow and extent of original stream gravel burial were considered). Until silted, these gravel beds could affect retention rates of these structures. From comparison of time-lapse photos, it appears that the three main aquifers were completely buried and thus filled up without groundwater transmission and volume reduction.

MONITORING NEEDED |

Green River
(medium to high gradient slopes)

Braided broad channels forming on low gradient reaches; drainage basin other than upper Green had minimal blowdown trees and singed trees, vegetative matter loss low, native soil relatively undisturbed. Volcanic ashfall deposits (upper reaches) forming alluvial fans in mainstream of channels. Channel and gully erosion developing in ashfall and alluvial fan materials, severely in clear-cuts. Headcutting to 1–1.5 m (3–5 ft) occurring in upper basin drainages. Morphologic changes few in lower basin, many in upper basin.

Restraining Structure,
Corps of Engineers (NFT)
(medium slope gradient; reaches upstream to Bear Creek; structural integrity of spillways critical to preserve storage capacity and retention of debris; see Table 3.2)

Flows occurring from Hoffstadt, Bear, North Fork Toutle on north, and North Fork Toutle on south sides. North side has sinuous channels and gullies incising through alluvium from upper reaches. Braided channels and gullies forming in low gradient reaches. Seeded grasses on hummocky volcanic pyroclastic deposit midvalley, stabilizing large sediment transportation and movement. South side has low gradient and large volumes of pyroclastic and alluvium material deposits in channel bed, incised deeply by headcutting gullies and channels. Drainage pattern mostly braided channels with numerous midstream sand and gravel bars. Many depressions, hollows, and craterlets exist in hummocky deposits. Present problem of this storage and retention area is the loss of the North Side spillway and removal of inflowing sediment (May 1982). Without adequate funding from federal sources, retention of sediment, filtering of large debris, and settling of silt and sand is lost. Structure was breached 3/19/82 and requires repair.

MONITORING AND DREDGING NEEDED

Flood Plains
(all rivers in Toutle River basin)

Major processes are formations of rills, gullies, and lateral incising of gullies into channels on the flood plains as feeder creeks and streams erode through poorly consolidated and only slightly cohesively bound flood deposits. Large volumes of flood-plain alluvium are being swept downstream into the Cowlitz and Columbia rivers, after passing through the Toutle River channels and forming flood deposits. This process is important during intense, heavy rainfalls, as peak flows move in the upper drainage basin. The rivers' discharges and velocities are increased to high levels; waters are very turbid.

aReferences: flight dates—9/9/80, 10/2/80, 10/28/80, 2/4/81, 6/16/81, 9/16/81, 3/5/82, and 5/23/82. Ground reconnaissance dates—9/28/80. Headcutting to 1–1.5 m (3–5 ft) occurring in upper basin drainages. Morphologic changes few in lower basin, many in upper basin.

TABLE 3.2
TOUTLE RIVER DRAINAGE BASIN DATA SUMMARY[a]

Reservoir or drainage basin	Size (sq miles)	Structural overtopping height (ft)	Major slope gradient range (%)	Capacity inflow (Ac-ft × 10³)[b]	Annual runoff (avg.) (Ac-ft × 10³)	Annual sediment yield (Ac-ft)	Average peak runoff (csm × 10³)[c]
Rivers							
North Fork Toutle (on Forest)	50 (16.34)	—	9	300	97.6	70,000 7,700	1.13
Green River Boundary (on Forest)	38.37	—	30–70	—	112.5	665	0.59
South Fork Toutle (on Forest)	50 (6.47)	—	10	—	31.1	12.5	1.13
Kalama River[d] (Unaffected)	8.34	—	10	—	44.3	—	0.60
Lakes and Reservoirs							
Spirit Lake	16.13	158	70–80	440/76	77.4	465	1.13
Coldwater and South Coldwater[e]	11.77 5.57	180	60–70	171/1.7	43.7	1,700	1.06
Castle Creek (South Fork)	3.10	90	60–70	22/14	14.9	300	1.13

North Fork Toutle Restraining structure (U.S. Corps of Engineers) 37 miles upstream from mouth of Toutle River (6100 ft × 35 ft high)	—	—	—	3700	—
Eight Sediment Basins, Lower Reaches of North and South Fork Toutle and Toutle River			(Data not available.)		

[a]Data taken from "Mount St. Helens Emergency Watershed Rehabilitation Report," Watershed Rehabilitation Team Members, U.S. Forest Service, Pacific NW Region, Portland, Oregon, 1980.

[b]Ac-ft ($\times$ 10^3) = 1.23 km^2/m.

[c]Csm ($\times$ 10^3) = 28.32 m^3/sec.

[d]Presented as an unaffected river in the area for comparison.

[e]South Coldwater Creek drainage rill erosion measured 7/10/80, >30% slope ranged from 57 to 144 tons/acre and in another transect on a road, 540 tons/acre.

4

Vegetative Succession Following Glacial and Volcanic Disturbances in the Cascade Mountain Range of Washington, U.S.A.

A. B. Adams and Virginia H. Dale

ABSTRACT

Similarities and differences in vascular plant succession in the Cascade range of Washington State are compared following the volcanic eruptions of Mount St. Helens, the Kautz Creek mudflow of Mount Rainier, and glacial recession in the upper canyon of the North Fork of the Nooksack River. Similar patterns of plant succession after volcanic and glacial cataclysms are the result of a common seed source from plants distributed widely by the common prevalent influences of these areas. These common environmental factors include a maritime climate, comparable mechanisms by which plants respond to localized disturbances, soil nutrients, soil particle size, and change in soil acidity with time. Differences in successional patterns relate to the occurrence of surviving plants, temporal heterogeneity associated with a chronosequence, microclimatic effects, the number and diversity of seeds present, and soil pH and salinity.

INTRODUCTION

Although geologists have described how volcanoes and glaciers are interrelated (Porter 1981, Virogradov 1981, Grosvald and Glazovskii 1981), ecologists studying plant succession after volcanic eruptions or glacial recessions (Griggs 1917, Smathers and Mueller-Dombois 1974, Manko 1974, 1975, Brink 1959, Stork 1963, Goldthwait et al. 1966, Viereck 1966, Reiners et al. 1971) have never related vegetation to the soil dynamics occurring after these large physical disturbances. In the Pacific Northwest, volcanism has produced sufficient topographic relief for the formation of glaciers. Volcanoes may actually cause ice advances by placing particulate matter into the atmosphere and

Surely this is too sweeping

70

increasing its albedo (Porter 1981). Large mudflows are triggered by heavy rains and the sudden release of water from reservoirs created by glaciers. Also, volcanic eruptions may cause mudflows by creating avalanches that displace large quantities of water out of lakes, by rapidly melting glaciers on their slopes, and by initiating rain via thermal convection. Mudflows associated with either glacial or volcanic activity are often composed of material produced by both disturbances (e.g., ash, pumice, and glacial till).

This paper will compare vascular plant succession in the Pacific Northwest following volcanic eruptions, glacial retreats, and associated localized disturbances. Vegetation response following the recent volcanic eruption of Mount St. Helens is compared with that found in an area of contemporary glacial retreat. Comparisons among patterns of plant succession are based on preliminary soil analyses, field observations, and greenhouse and field experiments. The areas considered have similar climates (Table 4.1).

The Pacific Northwest has several contemporary examples of plant succession following volcanic disturbances and glacial recessions (Fig. 4.1). The effects on vegetation after volcanic disturbances are exemplified by the May 18, 1980, eruption of Mount St. Helens (Mack 1981, Means et al. 1982, Adams and Adams 1982). Volcanic perturbations resulting from that eruption included a massive debris slide, an explosive laterally directed air surge, many mudflows, and several pyroclastic flows (Rosenfeld 1980). Smaller ash eruptions and pyroclastic flows also occurred on May 18 and 25, June 12, July 22, Aug. 7, and Oct. 16–18, 1980 (Lipman and Mullineaux 1981). A localized mudflow covered parts of the debris avalanche and North Fork Toutle mudflow on March 19, 1982.

The upper canyon of the Nooksack River in northwest Washington provides an excellent example of plant succession following a glacial retreat (Oliver and Adams 1979, Oliver et al. 1985). Materials accumulating on the floor of this canyon are being removed from a double cirque and the canyon walls; they are subsequently deposited by active glacial, alluvial, and colluvial processes. Although granodiorite is the major bedrock material, the moraines are predominantly phyllite derived from older metamorphic rocks exposed by glacial processes within the cirque. Fill is deposited as far as 1.8 km downstream from the terminus of the glacier, and outwash extends farther downstream (Bardo 1980). A chronosequence of recessions and advances occurred during the Neoglaciation period (Porter and Denton 1967, Denton and Porter 1970). In addition to the Nooksack Canyon, successional patterns on the Kautz mudflow (Frehner 1957, Nelson 1958, Ballard

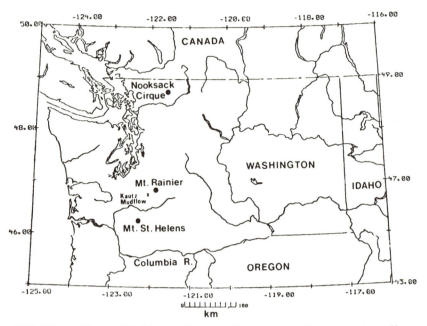

FIG. 4.1   Sites in Washington State used to compare plant succession effects of volcanoes and glaciers.

1963, Rigney et al. 1982) formed on the slope of Mount Rainier in 1947 are also described, and reference is made to the 350–500-yr-old pyroclastic flow down the west side of Mount St. Helens at Goat Marsh (Franklin et al. 1972).

    In this paper, patterns of plant recovery are compared following relatively common disturbances in the Pacific Northwest. The objectives of the study are to quantify the following: the site characteristics, the effect of soil conditions on plant growth, the phenological differences among species, and the influence of local disturbances on plant survival and establishment. Similarities and differences in plant succession following volcanic eruptions and glacial retreat in the Pacific Northwest are indicated. The study increases understanding of successional processes by documenting similarities in plant recovery within a geographic region. Experimental results described herein support a deemphasis on facilitation (Clements 1916, 1936) as a mechanism of plant succession because species characteristic of late successional stages germinate and grow in the new substrates produced by the eruptions of Mount St. Helens without being preceded by more opportunistic species. Nevertheless, particular circumstances exist in which allogenic and all types of autogenic succession (Dansereau 1954, Egler 1954) seem to be occurring. The patterns of plant recovery support the mosaic phenomenon of natural successions. Characteristics of the initial disturbance influence plant survival and site conditions. Seed and seedling traits, distance to the seed source, and soil properties all affect plant reestablishment. Localized disturbances superimpose a patchwork of seral stages and patterns on the developing community.

MATERIALS AND METHODS

SOIL ANALYSIS

Soil samples were collected from various substrates created by the 1980 eruption of Mount St. Helens, from a 350−450-yr-old pyroclastic flow at Goat Marsh on Mount St. Helens, and from parent material of the Kautz mudflow on Mount Rainier (Fig. 4.2). Collections were made at the surface and 15 cm below the surface. Particle size was determined by sieving for 15 min in a mechanical shaker for particles larger than 1/16 mm (14−250 mesh). For smaller particles a sedimentation technique was used. A wetting agent (sodium hexametaphosphate) was added to the samples to disperse the sediment into individual particles and prevent them from coagulating during subsequent analysis. Twenty-five grams of sample were analyzed because hydrometer readings indicated that the material was greater than 50% sand. The Wentworth scale was used for statistical manipulation and transformation ($\varphi = -\log_2 E$, where E is the diameter of the particle in millimeters). Since particle sizes were not normally distributed (Adams and Adams 1982), Folk−Ward values (Creager et al. 1962) were used to test for differences in heterogeneity, skewness, and kurtosis (shape of the distribution). To determine overall texture, Shepard values (Creager et al. 1962) were computed using the sand-silt-clay ratio.

Soil reactivity was determined by immersing a platinum electrode of a pH meter into the saturated soil paste. Water was extracted from the saturated soil pastes using a Buchner funnel, and the resistance of the filtrate was determined with parallel electrodes. Oven-dried soil was ground and digested using the micro-Kjeldahl procedure to solubilize nitrogen (Parkinson and Allen 1975). Moisture release curves were generated by placing soil samples within rubber O-rings on confined ceramic pressure plates (Richards 1965). The soils were saturated with water for 12 hr and then subjected to pressure (0.33−15 bars) with N_2 gas. Comparative soil properties of the Nooksack Canyon

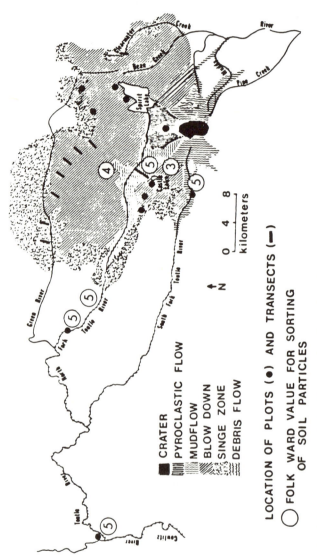

LOCATION OF PLOTS (●) AND TRANSECTS (▬)

○ FOLK WARD VALUE FOR SORTING
OF SOIL PARTICLES

FIG. 4.2   Location of the Devastation Zone in relationship to Mount St. Helens and the Cowlitz
Rivers. Sample plots (25 m²) were placed at 50-m intervals along linear transects (represented by black
bars). In 1981 25-m² plots were placed from Castle Lake to Coldwater Lake. Solid black circles portray
sites where species were surveyed over larger areas (3000 m²), and a minimum of three plots were
established. Circled numbers are Folk–Ward values for distributional differences in heterogeneity of
particles. The distribution of ash particles is more leptokurtic (better sorted) than the distribution of
particles in the more heterogeneous mud. All soil samples had Shepard values of 1 or 2, indicating sand
to silty-sand sized particles, with the ash typically being sand.

and Kautz mudflow were obtained from Bardo (1980), Frehner (1957), Nelson (1958), Ballard (1963), and Rigney et al. (1982).

GREENHOUSE STUDIES

To determine the availability of some elements, Romaine lettuce (*Lactuca sativa*) seeds were germinated in sterile sand and transplanted into 13-cm pots on April 15, 1981 (3 soils, 6 treatments with 4 replicates). Mudflow material was collected 2 km downstream from the terminus of the debris slide on the North Fork of the Toutle River (28 km from the crater of Mount St. Helens) and at the confluence of the Toutle and Cowlitz rivers (50 km from the crater). Debris slide material was collected north of Castle Lake (7 km northwest from the crater). Soil was filtered through a 0.64-cm screen, and 1200 g was added to each pot. One treatment with no added nutrients served as a control. The complete macronutrient treatment included nitrogen, phosphorus, and potassium, and partial treatments omitted one nutrient at a time (Appendix A). One treatment had complete macronutrients as well as micronutrients. After 4 weeks the above-ground biomass was harvested, dried at 70°C, and weighed. The data were subjected to analysis of variance to determine differences among soils, nutrient additions, and replicates. Since absolute yields can vary with season, the relative yield was calculated.

To assess seedling characteristics, seeds of native tree species found on the Kautz mudflow (Frehner 1957) and Mount St. Helens before the eruption (Kruckeberg 1980 and Kruckeberg, Chapter 2, this volume) were planted in 8-cm-deep flats filled with mudflow material from the North Fork Toutle. Prior to planting, the seeds were weighed and conifer seeds were stratified for 4 weeks. The seedlings were kept in the greenhouse for 5 months, and then the fresh and dry weights of roots and above-ground tissues were determined.

To examine the possibility that certain species could survive as small plants on a mudflow, saplings of eight native tree species were planted in moisture stress troughs (Pickett and Bazzaz 1976) 0.54 m³ in volume. The troughs were lined with plastic, placed on a 7° slope with a southerly aspect, partitioned with a water permeable barrier into six equal volume compartments, and filled with mud from the North Fork Toutle. Deciduous trees in troughs were maintained in a greenhouse under 16 hr of light. *Alnus rubra* seedlings were germinated from seeds collected from a tree near Mount St. Helens. *Salix sitchensis* and *S. scouleriana* were cloned from trees growing near Kent, Washington, on a lahar deposit (the Osceola mudflow) that originated on Mount Rainier over 5000 yr ago (Crandell and Waldron 1956). Each trough of *Salix* contained one genotype. The troughs with conifers (*Pseudotsuga menziesii, Abies amabilis, A. procera, Tsuga heterophylla*, and *Thuja plicata*) were kept outside the greenhouse at the University of Washington in Seattle. Leaf conductance was determined according to Turner and Parlange (1970).

To test the hypothesis that mean height varies as a function of soil moisture content, the change in height after 7 months was subjected to analysis of variance and Scheffe's contrast test. All differences reported are for $p > 0.05$. Colwell's nutrient solution (Colwell 1943) was added to the soil in one trough containing *Alnus* and four troughs with male and female rootings of *Salix scouleriana* and *S. sitchensis*. Each plant received 200 ml of solution per month for the first 4 months. The nutrient formula contained all the major elements required by plants without causing excessive buildup of trace minerals. Trace elements included manganese, copper, zinc, and boron.

FIELD METHODS

Field studies in the uppermost canyon of the North Fork of the Nooksack Canyon (Fig. 4.3) were conducted in mid-July to early September 1977 (Oliver and Adams 1979). Percentage of cover was recorded for each species in 177 circular plots of 40 m² spaced at 50-m intervals along linear transects through areas of homogeneous physical structure. Communities were defined using a polythetic, hierarchial, agglomerative clustering strategy (Orloci 1969, Goldstein and Grigal 1972) in which groups are plotted on the basis of commonality, richness, and abundance of species. Directions and frequencies of snow slides were estimated by analyzing the tree rings for scars and changes in growth rates. The positions of moraines were correlated with ages determined from tree rings to set age class boundaries created by glacial movements. Diversity measures (richness and the Brillouin proportionality index [Peet 1974]) were calculated for each plot.

For the Mount St. Helens area, vegetation frequency and cover were measured during the summer of 1980 in 25-m² circular plots placed at 50-m intervals along linear transects (Fig. 4.2). On the debris slide, 250-m² circular plots were placed at 50-m intervals running north of Castle Lake in June 1981. Species lists were constructed during forays over larger areas (3000 m²). Seed traps of 0.25-m² square pieces of mosquito netting smeared with sticky oleoresin were placed 0.5–1.0 m above the ground surface at the edge of the debris slide at Castle Lake and in the blowdown area east of Spud Mountain in 1980. In 1981, 30 seed traps were placed within 250-m² plots in the transect

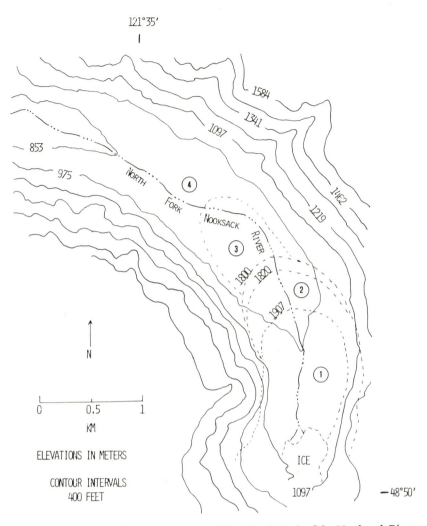

FIG. 4.3 Map of the upper canyon of the North Fork of the Nooksack River located on the border of North Cascades National Park in northwestern Washington. Large circles with numbers indicate age classes, and the dotted lines delimit the age classes. Age class 1 and 2 are well defined by the 1907 and 1800 moraines and are less than 75 and 180 yr old, respectively. Age class 3 is 300–350 yr old, and the oldest sites in age class 4 have not been glaciated for at least 600–800 yr and probably not since the Pleistocene.

running north–south from Castle Lake. In 1982, 44 traps were placed at 50-m intervals along a west–east transect on the debris slide from the border of the national monument to below Elk Rock. Seeds of five conifers (*Thuja plicata, Pseudotsuga menziesii, Tsuga heterophylla, Abies amabilis, A. procera,* and *Pinus contorta*) and two deciduous trees (*Populus trichocarpa* and *Alnus rubra*) were planted at a density of 200/0.5 m² on the debris slide and ash. Small conifer saplings of the same five species (20 trees per species) were planted in ash surrounding Castle Lake. The work by Hitchcock et al. (1973) was used as the taxonomic authority for vascular plants.

RESULTS

SOIL ANALYSES

Particle size analyses for fractions >2 mm demonstrate that all samples have Shepard values of 1 and 2, indicating sand or silty sand (Table 4.2). Mudflow samples are roughly 20% silt and 3–5% clay, whereas debris and ash samples are more variable and difficult to characterize. Ash collected after the May 18 eruption is almost entirely sand, except for collections from Elk Rock. This may reflect that sites northwest of the volcano were affected more by the earthquake-triggered blast in May than by subsequent ashfalls (Decker and Decker 1981). The mud is more heterogeneous than the ash (Folk–Ward values of 5 for sortedness compared to 3 and 4 for ash). The particle size distribution of mud is more positively skewed, while kurtosis is variable from sample to sample. Debris slide material collected in July 1980 at a depth of 15 cm has more silt (Figs. 4.4 and 4.5).

The equations of best fit for moisture release from soils are given in Table 4.3. Grams of H_2O per gram of soil (Y) is related to pressure exerted by N_2 gas (Ψ) by the expression $Y = a\Psi^{-b}$ (Figs. 4.6–4.8), where a is the intercept and b is the slope. Since the proportion of small rocks influences the ability of soil to retain water, the percentage of rocks is presented in Table 4.3. The >2 mm fractions and rocks from sites around Mount St. Helens are shown in Fig. 4.9.

Mud and debris slide material from the Mount St. Helens area are compared with material collected from the Kautz mudflow and the Nooksack glacial area. Initial pH values are neutral for mud and ash (Tables 4.3 and 4.4), but are slightly acidic for debris slide and most pyroclastic samples. Two pyroclastic samples collected next to a sulfatero, however, have the lowest pH values by two orders of magnitude

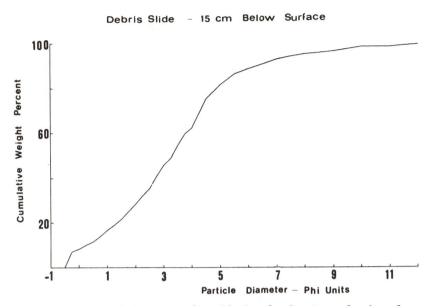

FIG. 4.4   Cumulative curve of particle sizes for the <2 mm fraction of one soil type used in the Jenny pot test taken from the debris slide 15 cm below the surface. This curve is identical to that obtained for the mudflow material collected from a site 2 km below the terminus of the debris slide from Mount St. Helens at the confluence of the Cowlitz and Toutle rivers.

(see sulfatero in Table 4.3). Electrical conductivity values are generally less than two, indicating low salt concentrations, but substrates collected adjacent to active fumaroles have high conductivities and thus high concentrations of salts. Comparing pH and conductivity values of the 350–450-yr-old pyroclastic material of Goat Marsh with depositions of the 1980 eruptions suggests that the older material is most similar to the debris slide and nonsulfatero pyroclastic material.

All values for nitrogen are low but nevertheless greater than values reported for the Kautz and Nooksack during the first years of deposition (Table 4.4). Carbon and nitrogen are highly variable among the samples, ranging from 350 to 2960 ppm for carbon and 24.5 to 703.0 ppm for nitrogen. As a result, the carbon to nitrogen (C/N) ratios reflect this variability with ranges from 0.23 to 75. Nitrogen is lowest in material from the Nooksack, which also has the highest C/N ratio. Material 15 cm below the surface of the debris slide has the greatest amount of nitrogen and the lowest C/N ratio.

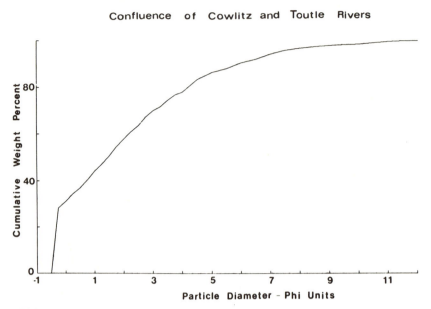

FIG. 4.5   Cumulative curve of particle sizes for the <2 mm fraction of another soil type used in the Jenny pot test collected from a site on the debris slide just north of Castle Lake near the confluence of the Cowlitz and Toutle rivers at a depth of 15 cm in July 1980. The debris slide material has less slope than the mudflow substrate due to the presence of more silt.

GREENHOUSE STUDIES

Results of the Jenny pot test demonstrate significant growth differences between lettuce grown in mud and debris substrates produced by the May 18, 1980, eruption (Table 4.5). The analysis of variance shows statistical differences between the two mean lettuce weights based on soils and nutrient additions. There is no difference in lettuce growth between soil treatments without the addition of nutrients or the addition of phosphorus or nitrogen by themselves. There is a significant difference in lettuce response between the soil of the upper debris slide and soil from mudflow material with the addition of phosphorus and nitrogen together. The farther the material is collected from Mount St. Helens, the greater the biomass of the lettuce. Both the debris and mudflow materials are depleted in nitrogen and phosphorus (Table 4.6). A field response is obtained with a relative yield for lettuce of 30% for nitrogen and 20% or less for phosphorus (Jenny et al. 1950). These criteria are met for both nutrient treatments. The deletion of potas-

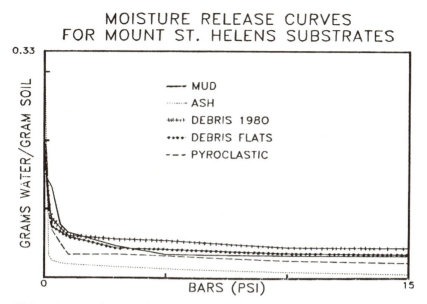

FIG. 4.6   Moisture release curves for four soil types from the Mount St. Helens area. Values at zero pressure are field capacities. A locally weighted, scatter-plot smoothing technique (Cleveland 1979) was used to generate curves (see Table 4.3 for parameter values). A. Ash collected from Spud Mountain, August 7, 1980 ($N = 27$). B. Nonsulfatero pyroclastic material collected 0.2 km north of the lava dome on May 12, 1980 ($N = 33$). C. Mudflow substrate collected below the debris slide and at the confluence of the Toutle and Cowlitz rivers in July 1980 ($N = 28$); mud represents material used in the moisture stress troughs and the Jenny pot test. D. Debris slide substrate collected from both flats and mounds in July 1980 and August 1981 ($N = 61$); material from debris slide flats is the same as was used in the Jenny pot test.

sium from the nutrients results in an increase in biomass of the lettuce plants grown in the mudflow material. Doubling the amount of phosphorus in the nutrients does not significantly increase lettuce growth relative to the complete nutrient treatment. Poor growth is achieved in the debris slide substrate with all treatments and with the addition of micronutrients to the complete treatment, suggesting a possible toxic effect or a structural limitation of that soil.

Comparisons of seed and seedling traits of seven native trees may explain differences in success and successional status on mudflow deposits (Table 4.7). All species have the ability to germinate and survive on the material, but *Alnus rubra* and *Pseudotsuga menziesii* have the greatest seedling weight compared to seed weight. *Alnus rubra*

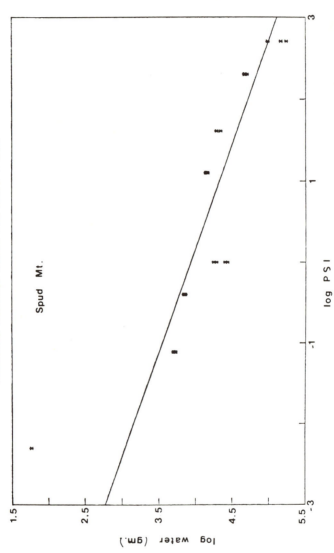

MOISTURE RELEASE CURVE FOR AUG. 7 ASH

Spud Mt.

log water (gm.)

log P S I

FIG. 4.7   Log–log transformation of moisture release curve for Mount St. Helens soil. Soil samples collected at the confluence of the Cowlitz and Toutle rivers in July 1980 were used in the moisture stress troughs and the Jenny pot tests. Curve indicates that mud releases water more gradually and over a greater range than other substrates (see Fig. 4.8).

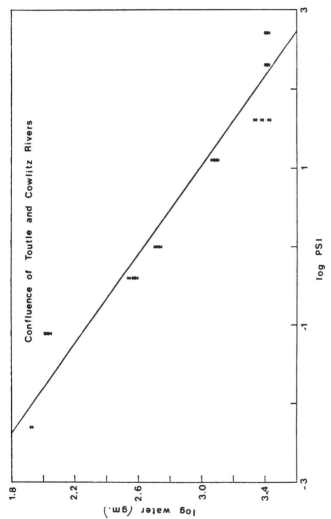

FIG. 4.8   Log–log transformation of moisture release curve for soil collected at Spud Mountain from the August 7, 1980, ash fall. Curve demonstrates that the ash holds more water at low values of ψ because of the pores in the coarser ash particles, but that the ash loses this water rapidly because of the lack of fine silt and clay particles (see Table 4.2). Pores in the ash explain the large residuals for values < -0.1 bar.

Pyroclastic

24.79%
rocks

North of Crater

A

Ash

0.001%
rocks

Spud Mountain

B

FIG. 4.9   Four soil samples from the Mount St. Helens area illustrating the
<2 mm fraction, the >2 mm fraction, and representative rocks (Table 4.3 gives
the relative weight of each component). The soils were collected from non-

sulfatero pyroclastic deposits (A) just north of the crater, Spud Mountain ash (B), debris mounds (C) at the northwest corner of Spirit Lake, and the Jackson Lake debris mounds (D).

seedlings were nodulated 5 months after germinating. *Pinus contorta* had the highest germination success but did not grow as fast as *A. rubra* or *P. menziesii*. Both species of *Abies* tested have heavy seeds and low germination rates. *Thuja plicata* and *Tsuga heterophylla* have the lightest seeds of the conifers and have seedling traits intermediate to those of the other species tested.

Leaf conductance measurements indicate that all tree saplings planted in moisture stress troughs respond to different moisture levels $[\ln (Y + 1) = -0.11 + 0.24X, r^2 = 0.83, p < 0.05$ for *Salix sitchensis*, where Y is the position in the trough varying from wettest to driest and X is leaf conductance measured in cm/sec^2]. Growth rates of native trees in mud suggest that trees of all species can survive when planted directly in the material without fertilization or preconditioning by other species. There were, however, significant differences between species and moisture levels. *Salix sitchensis* in troughs to which fertilizer was added grew more than any of the other species (as tall as 110 cm) and had a significantly greater increase in height than *S. sitchensis* in troughs to which no fertilizer was added. *Salix scouleriana* grew the least with all treatments. *Salix* that was transplanted into troughs to which no fertilizer was added produced new leaves only in the wetter habitats. *Alnus rubra* in fertilized and unfertilized mud had similar increases in height, with the greatest height attained at soil water potentials of -0.15 to -0.20 Ψ (corrected for rocks). Mortality was greater for *A. rubra* in the unfertilized mud (40%) than in fertilized mud (8%). Variation in growth of *A. rubra* was great in fertilized mud, suggesting that the success of it on mudflows may be genetically variable from individual to individual and may be closely related to nutrients. To check for establishment of nitrogen-fixing root symbionts (Trappe et al. 1968), five *A. rubra* were removed from mud to which no fertilizer had been added and nodulation was detected on roots in all cases.

Of the conifers in the troughs, only saplings of *Pseudotsuga menziesii* grew at all moisture levels. Individuals of this species retained their leaves during the winter and broke bud in early April. *Abies procera* grew the poorest of all of the conifer transplants. *Abies procera*, *A. amabilis*, and *T. heterophylla* died on waterlogged mud. Although these individuals lost some of their foliage in the fall, much of the mortality occurred during a 2-day period in which the waterlogged soil froze.

FIELD STUDIES

The frequency of occurrence of each species found within the blast zone during the first growing season following the May 18 eruption is given in Appendix B by type of physical disturbance in which the species were found. Also indicated are species in the youngest age classes of the Nooksack Canyon and noted by Frehner (1957) on the Kautz mudflow. Listed are 78 common species for the Mount St. Helens area, 71 for the three youngest communities at the Nooksack Canyon, and 48 for the Kautz mudflow by Frehner (1957). In this list, 10 species are shown as common to Mount St. Helens, the Kautz mudflow, and Nooksack Canyon. A species list compiled for the devastation zone in 1982 increased the number of species found to 226 (Appendix C). Comparison of this list with Appendix B indicates 14 species common to all three sites. A total of 40 genera and 37 species are common to the youngest age class of the Nooksack area and the area affected by the lateral blast when compared with the 1982 species list. In the blowdown and singe areas, richness is related to the depth of ash deposited and the type of volcanic disturbance to which a site was exposed (unpublished data). Most of the species found within the devastation zone listed in Appendix C represent survivors. By 1984 the list had increased to 249 species.

Residual plants found in the devastation zone of Mount St. Helens (Adams and Adams 1982) and species from the Nooksack Canyon and the Kautz mudflow were categorized by life form as delimited by Raunkiaer (1934) and refined by Mueller-Dombois and Ellenberg (1974) (Table 4.8). Habitats after the Mount St. Helens eruption were dominated by geophytes with dormant buds located below the ground surface. There, however, differences between the type of physical disturbance, species cover, and frequency (Adams and Adams 1982). Similarly, in the Nooksack Canyon geophytes were predominant and small shrubs (chamaephytes) and annuals were rare. The percentage of cover of chamaephytes was high on eroded till and moraines in the Nooksack, even though the proportion of species was low. Unlike the newly created Mount St. Helens habitats, the Nooksack area had more species with buds at the surface and the Kautz mudflow had a rapid recruitment of geophytes and trees.

In 1980 seed traps placed on the border of the debris slide and blowdown areas caught seeds of three vascular plant genera (*Agrostis*, *Epilobium*, and *Cirsium*) and lichen fragments (*Cladonia* sp. and *Alectoria sarmentosa*). Seedlings of *Epilobium* and *Cirsium* were found on the debris slide and in the blowdown area. During July 1980, four seedlings of *Tsuga mertensiana* were found east of Spud Mountain, and a *Populus trichocarpa* seedling was found above Schultz Creek (~3 and 7 km from the nearest conspecifics, respectively); both sites were located in the blowdown area. Trees surviving for a short time on mudflows could act as seed sources (Frehner 1957, Beardsley and Cannon 1930).

Standing *P. trichocarpa* set seed after the May 18 eruption, and in mid-June more than 400,000 seedlings/ha were found on the mudflow between the western edge of the debris slide and the confluence of the Cowlitz and Toutle rivers. On mudflows and in blowdown areas, germination and survivorship was restricted to sites with relatively little temperature and moisture fluctuation (e.g., depressions). Frehner (1957) reported frequency for particular species (notably of the family Asteraceae) to be greater under snags on the Kautz mudflow.

The germination and survivorship of eight species in field seed plots was highly variable (Table 4.9). Among individual plots, some had no successful germinations, while others had as many as 11% of the seeds germinate and survive the first growing season. Many plots were completely washed away by the winter rains. The success of seedlings was greater for plots on the north side of snags than for those located in open areas. These experiments supported field observations made on the Kautz mudflow (Frehner 1957) that seedlings in more protected sites were more likely to survive than seedlings in the open. Eight conifer seedlings were found on the debris slide in 1981 at a density of 3/ha (Table 4.9). In 1982 conifer seedling density had increased to 40/ha. One *Alnus rubra*, several *Populus trichocarpa* and *Salix lasiandra*, and several thousand *S. scouleriana* and *S. sitchensis* were also growing on the debris slide in 1981. Although no plants were found on the pyroclastic flow north of the crater in 1980, by May 1981 moss gameto-phytes had become established and seedlings of 12 species were evident in the warmer areas located at the base of the "steps" (a series of bluffs north of the crater) (see Table 4.10 and Fig. 4.11). Many species blanketed by the blast and debris were uncovered by the winter rains of 1980–1981 and 1981–1982 and appeared as patches of green on the sides of deep gullies. Conifer plugs planted in October, 1980, showed a high rate of mortality. The surviving ones lost much of their foliage during the winter of 1980–1981, but did produce new leaves by May 1981. All the *Abies procera* and *Tsuga heterophylla* died, but 69% of *Pseudotsuga menziesii*, 8% of *Thuja plicata*, and 2% of *Abies amabilis* were still alive after 1 yr.

Seed and seedling data collected in 1981 and 1982 showed an abundance of seeds representing many growth forms being recruited onto the debris slide (Table 4.10 and Figs. 4.10, 4.11, and 4.12). The most abundant types found both as seedlings and seeds were wind-dispersed seeds with plumes (e.g., *Epilobium angustifolium*, *Cirsium* spp., and *Salix* spp.). Less abundant were wind-dispersed seeds with wing-like structures such as *Carex* spp. and *Alnus rubra*. Animal-dispersed species found on the 1981 debris slide included *Rubus ursinus* and *Vaccinium membranaceum*. In 1981 the greatest number of seeds were

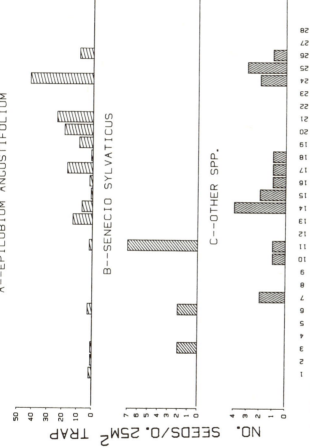

FIG. 4.10 Number of seeds per trap versus position along the transect running south–north from Castle Lake to Coldwater Lake across the debris slide (see Fig. 4.2). The first two histograms (A and B) are for the most abundant seeds captured, whereas the third (C) represents all other seeds captured, including those of *Cirsium* and *Sonchus*. Data were collected from August 24 to September 3, 1981.

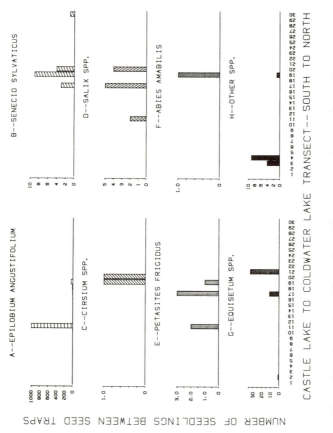

FIG. 4.11   Number of seedlings found between seed traps along the Castle Lake to Coldwater Lake debris slide transect (see Fig. 4.2). A–C. Histograms of species whose seeds were also caught in seed traps during the same time period (Fig. 4.10A–C). D–G. Histograms of species not detected in seed traps. H. A histogram of all other seedlings detected (*Anaphalis margaritacea*, *Carex* spp., *Festuca* sp., *Hypochaeris radicata*, and *Rubus ursinus*). Data were collected on August 24, 1981.

those of *E. angustifolium* followed by *Senecio sylvaticus*. Similarly, these two species represented the largest number of individuals found in a count of seedlings found along the transect running from Castle Lake to Coldwater, but *S. sylvaticus* was fifth in numbers found in plots on the debris slide (Table 4.9). Although seedlings of *Petasites frigidus* and *Salix* spp. were found throughout the debris slide in riparian areas, no seeds of these taxa were collected since they produce seed in the spring prior to the time during which our seed traps were set. For similar reasons, it was improbable for us to catch conifer seeds during the period sampled, and of course *Equisetum* spp. disperse by spores.

A notable seasonal trend was detected in seed dispersal with both a north–south transect and a west–east transect. In addition, position along the transect seemed to be an important factor (Fig. 4.12). Both seeds and seedlings had higher densities in the center of the debris slide in 1981 than at the borders (Figs. 4.10 and 4.11). This result was reproduced in 1982 (Fig. 4.12A). As shown with time-series analysis of the Castle Lake to Coldwater Lake transect, the number of seeds peaked in the middle of the debris slide in late summer. Nevertheless, a relatively higher number of seeds were caught on the borders of this transect than in the center during July and August. Along the west–east transect there was little variability from June to August, and few seeds were caught. Yet, in late summer and early fall many plumed seeds were caught and these initially showed higher abundance on the western end of the transect nearest to the border of the blast zone. A relative increase in seeds in the upper Toutle valley was detected with a shift in seed abundance from the western end eastward along the transect in late September and early October (Fig. 4.12B, bottom graph, two subseries on the right).

Below the Nooksack glacier *Abies amabilis*. *Tsuga heterophylla*, *T. mertensiana*. *Pinus contorta*, *P. monticola*, *Thuja plicata*, *Chamaecyparis nootkatensis*, and *Alnus sinuata* occupied sites within a few years after glacial recession. In an area exposed by the glacier for >37 yr, 682.4 overstory stems per hectare occurred (37% *Tsuga mertensiana*, 19% *Abies amabilis*, 42% *T. heterophylla*, and 2% other species). The age of trees ranged from 1 to 37 yr with a maximum height of 23 m. Early recruitment of these late successional species supports the concept of initial floristics of Egler (1954) in that early establishment of late successional species is possible and that the direction taken and equilibria reached by plant communities may be determined by the initial colonizers. For sites at the Nooksack Canyon over 800 yr old, 22.1 overstory stems per hectare were recorded (3% *T. mertensiana*, 67% *A. amabilis*, 29% *T. heterophylla*, and 1% other species) (Oliver and Adams 1979,

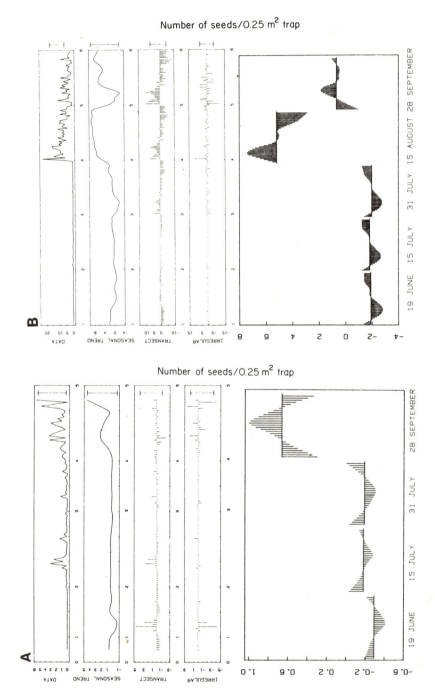

Number of seeds/0.25 m² trap

Oliver et al. 1985). Areas in the Nooksack that were subsequently deglaciated were inundated by snow avalanches.

The diversity of community types in the canyon could be related to the frequency of snow avalanches and wind throws (Fig. 4.13). Maximum diversity occurred at a disturbance frequency of once every 10 yr. With this disturbance frequency a patchy shrub community was maintained; there were at least 11 species of shrubs as well as small broken trees and herbs within a 40-m² plot. In contrast, below the glaciers the maximum number of species and distributional diversity occurred approximately 80 yr after glacial recession. For longer times, diversity decreased and then increased again in old growth forest as the canopy opened and recruitment in the understory was reinitiated.

As Matthews at Storbroca

DISCUSSION

DIFFERENT EFFECTS OF GLACIATION AND VOLCANISM ON PLANT SUCCESSION

Comparisons between glacial succession at the Nooksack Cirque, five habitats created by the May 18, 1980, eruption of Mount St. Helens, and the Kautz mudflow show an array of differences. The effects of

FIG. 4.12   Time series analyses of seed dispersal for the Castle Lake to Coldwater Lake debris slide transect for four time periods with 30 seed traps (A) and the west−east debris slide transect for five time periods with 44 seed traps (B). The dates refer to the middate for the time interval of each transect subseries. The top four panels for each transect represent the actual data (top panel) and the three components (seasonal, transect, and irregular). The seasonal component represents the long-term trend associated with changes in seed dispersal throughout the summer and early fall. The transect component represents variability that may be accounted for by the position of a given seed trap along a transect. The irregular component is that noise left over when variability has been subtracted for the seasonal trend and position of a given trap along a transect. The lower graph is a seasonal plot with position along the transect as a subseries: first, the June values are plotted for successive seed traps along the transects, then for July, and so forth. For each transect subseries, the midmean of the values is portrayed by a horizontal line. The values of the traps along the transect are then portrayed by a vertical line emanating from the midmean line. The scales of the panels are not the same. The bars to the right portray the relative scaling by representing the same amount of change in the data and components. The decompositions were run with the length of the monthly trend smoother equal to 15 and the length of the transect smoother equal to 11 in both cases (Cleveland and Terpenning 1982). Both the seasonal and transect components account for substantial amounts of variation in both series. Data for both were collected in 1982.

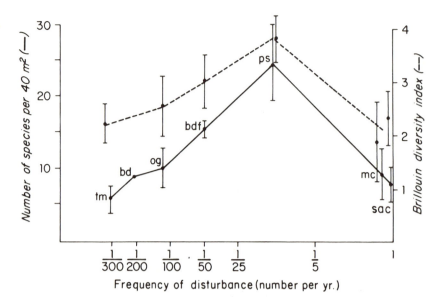

FIG. 4.13   Two measures of diversity plotted as a function of frequency of disturbance (snowslides or windthrows). Initials represent communities defined in Oliver and Adams (1979), as follows: tm = *Tsuga mertensiana* forest, bd = snowslides into *Tsuga mertensiana* forest at 200-yr intervals, og = windthrows into old-growth forests at 100-yr intervals, bdf = snowslides into forests at 50-yr intervals, ps = areas of patchy shrubs and broken trees, mc = maple chutes, and sac = south facing alder chutes. Bars represent ±1 S.D. Although species composition is different in the various communities, richness is significantly different only for ps (Duncan's Multiple range test, a = 0.05). For Brillouin indices, the diversity of mc is less than that of tm, og, sac, and bdf, respectively, which are in turn all significantly less than ps.

these major disturbance types on various aspects of plant succession may be subdivided into physical structure of the substrate, soil chemistry, community structure, plant recruitment, the presence or absence of a chronosequence, and climatic effects (Table 4.11).

Physical Structure of Substrates
Both the deposits in the Nooksack Canyon and at Mount St. Helens are variable, yet the substrates produced by the eruption of Mount St. Helens are more complex. At the Nooksack Canyon soil stratification is not prevalent in the youngest deposits. Rubble and boulders are intermixed with sites of eroded till and morainal mounds (Appendix B). Such substrate mosaics influence plant succession. For exam-

ple, pockets of fine till in outwashes, depressions, and behind large moraines support the first successful invaders (*Alnus sinuata, Juncus* spp., *Carex* spp., and many genera in the family Poaceae). These vegetation islands gradually expand (Sharitz 1973).

Most deposits within the Mount St. Helens devastation zone have a large proportion of sand-sized particles, and the surface of the ash is often covered by a thin layer of compact, finer ash. Aeration and drainage is more suitable for plants if they can first penetrate the upper layers of fine ash. The debris slide and slurry mudflows have much more heterogeneous substrates than other deposits at Mount St. Helens, but the mudflow components are less sorted than wind-sorted ash. Pyroclastic flows exhibit unique sorting of particles (as discussed below). Soil stratification is obvious within recent deposits of the Mount St. Helens devastation zone where strata of popcorn pumice and coarse and fine ash are easily resolved.

The uniform nature of pyroclastic material and its inability to retain water may restrict recruitment of plants to drought-tolerant species. These characteristics may explain why some eastern Cascade and fire-adapted species occurred on Mount St. Helens (e.g., *Arctostaphylos nevadensis* and *Pinus contorta*) and why high altitude plants (e.g., *Abies procera, Pinus albicaulis,* and *Lupinus lepidus* var. *lobbii*) are found at lower altitudes than normal on pyroclastics as well as on serpentine (Kruckeberg 1951, 1954, 1967). Although the moisture release curve for Mount St. Helens pyroclastic material is intermediate when compared to other substrates produced by the volcano, our field observations are consistent with the hypothesis that the slowest establishment of vegetation will occur on pyroclastic material.

Soil Chemistry
Soil reactivity influences nutrient uptake by plants. The pH of Nooksack substrates differs from Mount St. Helens and Kautz mudflow substrates by being initially more acidic (Bardo 1980). The low pH and high conductivity (an estimate of salinity) values reported for the sulfateros can inhibit the growth of some plant species (Richards 1954) and influence the structure of plant communities (Voroshilov and Sidel'nikov 1979). Some plants, however, have been shown to adapt to SO_2 (Winner and Mooney 1980). Properties of the soils alone or in conjunction with released atmospheric SO_2 may have been a factor controlling past vegetation patterns on Mount St. Helens (Lawrence 1954, Saint John 1976, Kruckeberg, Chapter 2, this volume).

The effect of salinity also is related to other physical properties of the substrate (e.g., particle size distribution and its relationship to soil moisture retention). If a process that produces a saline soil no longer exists, or if the salinity is created temporarily by some anomalous event, then changes in species composition through time can result only from this abiotic variation. In such a case the plants play a passive role. Changes in such communities are due to abatement of the initial source of salinity or to leaching of the salts. On the debris slide where water seeps to the surface and evaporates, a white saline crust forms. These salt efflorescences may influence plant establishment and community structure. Salt deposits were not observed at the Nooksack Canyon or at other areas in the Pacific Northwest with receding montane glaciers.

Large differences are found in the C/N ratios of mud and debris slide substrates produced during the May 18 eruption of Mount St. Helens, which suggest variable sources of nitrogen (e.g., roots, wood, leaves, and soil). Variation in the C/N ratio is not characteristic of young material at the toe of the Nooksack glacier (Bardo 1980). Pyroclastic material is noted for developing high C/N ratios (Ugolini and Zasoski 1979), but to our knowledge the initial variability described here for mudflows and debris slide material has not been previously reported.

The results of the Jenny pot tests with lettuce can be interpreted in two ways. The first is that the poor response to all nutrient additions by lettuce growing in debris slide material may be the result of limitations due to soil chemistry (e.g., nitrogen and phosphorus). The second possibility is supported by cumulative curves of particle sizes (Figs. 4.4 and 4.5) showing that debris has a great deal of fine silt. These smaller particles may become compacted and present a barrier to the growth of lettuce roots. Both soil chemistry and silt also may be interacting to limit lettuce growth. Natural substrates with small particle sizes may be more difficult for the establishment of seedlings. Substrates with coarser material can provide more aeration and offer less resistance to root growth. Once plants do become established, however, water availability will be greater with the finer substrates.

COMMUNITY STRUCTURE AND PLANT RECRUITMENT

After a cataclysm, organic remnants influence successional processes. In Mount St. Helens mudflows at lower elevations (1000 m or less), existing snags and logs influence the survival of early recruits as was observed on the Kautz mudflow (Frehner 1957). Also, trees that survive for a short time act as seed sources and influence the type of vegetation that is reestablished on the mudflows. The blowdown area has an abundance of logs and stumps that affect the return and survival of vegetation in many ways. Beneath logs, less ash accumulates, and it is at these locations that residual plants persevere and seedlings are surviving. Logs provide shelter and, ultimately, nutrients for plants.

A characteristic of weedy species, that is, early successional and opportunistic species (R-selected species in the sense of MacArthur and Wilson [1967]) is their ability to survive disturbances (Baker 1974). With the May 18 eruption of Mount St. Helens the survival mechanism was regeneration from *in situ* underground buds and transported fragments. Of the areas considered in this paper, the greatest number of species exist in the blowdown and singe area in the blast zone north of Mount St. Helens. These residual plants are most abundant in areas leeward to the directional blast, areas covered by snow on May 18, 1980 (e.g., north-facing slopes) and areas of moderate erosion subsequent to the eruption. Many of these plants (e.g., *Trillium ovatum*) regenerated from axillary buds, which have been shown to be activated on removal of the aerial shoot in early June, but not activated following removal of the shoot in late June (Kachura 1974). Many plants, however, are not able to produce new shoots during the season if the above-ground tissue is removed during the period of rapid growth that occurs in the spring. This is especially true of plants that lack subterranean buds. The scarcity of species such as *Berberis nervosa* and *Sambucus racemosa* in 1980 might have been due to the timing of the eruption. Had the eruption occurred later in the season, probably fewer individuals would have produced above-ground shoots in the summer of 1980, but the number of small shrubs with above-ground foliage might have been greater.

Although mudflows and the debris slide of Mount St. Helens do show immediate regeneration from plant fragments and thus have more species than barren glacial areas, species richness was higher on glacial substrates at the Nooksack after 35 yr than on the Kautz mud-flow after the same amount of time had elapsed (the areas are roughly equal in size). The areas of highest species diversity on the debris slide were islands of root wads and/or mineral soil that washed down with the avalanche and were emplaced close enough to the surface for plant fragments and/or seeds to sprout and penetrate the deposit surface, and areas on the debris slide where erosion rapidly exposed pockets of mineral soil containing surviving plant species.

As discussed previously, no residual species were found on the pyroclastic flows visited in 1980. Based on the vegetative analysis of the 350−450-yr-old pyroclastic flow above Goat Marsh (Franklin et al. 1972), few species would be expected to tolerate the adverse conditions on pyroclastic material.

The ability of plants to survive a catastrophe has significant effects on the rate of succession. Immediately following pyroclastic flows, lava flows, and glacial recessions, all residual plants and soil are gone. Succession is primary on these substrates. In contrast, mudflows, the debris slide, and singe and blowdown areas of Mount St. Helens are undergoing secondary succession because underground organs and organic material remain.

The distance from a seed source, the number of seeds produced by a species, and the dispersal capability of seeds will influence both the number and types of offspring found on newly exposed substrates (MacArthur and Wilson 1967, Platt 1975, Carkin et al. 1978, see fig. 6 of Frehner 1957). One might predict that pyroclastic and debris slide deposits at Mount St. Helens receive a less diverse seed immigration than the singe perimeter, blowdown and mudflows because of the increased distance from seed sources in the surrounding forest. The blowdown and singe areas of Mount St. Helens are also heavily invaded by ruderal and fugitive species (Hutchinson 1951) that survived the lateral blast. Surprisingly, conifers have become established at an increasing rate on the debris slide.

Establishment of trees is more difficult because their seed has the greatest distance to travel from source to site of disturbance. Since mudflows and singe areas of Mount St. Helens are bordered by standing forests, they have a closer, more diverse seed source. Areas left bare after retreats by montane glaciers such as the lower Nooksack glacier have a diverse seed source because of the proximity of living plants, not only downstream from glacier but also from tree clumps and subalpine meadows situated above recently deglaciated areas. Plant species that disperse readily by seed have demonstrated their ability to grow close to glaciers (Oliver et al. 1985).

CHRONOSEQUENCE AND CLIMATIC EFFECTS

Long chronosequences are common to glacial recessions but are much shorter for the volcanic disturbances discussed here. For example, at the Nooksack Canyon the vegetation is progressively younger closer to the glacier. Mudflows, the debris slide, and the pyroclastic flows of Mount St. Helens were deposited in a shorter time relative to an extended glacial retreat such as at the Nooksack. As a result, the noninteractive phase of plant recruitment during which seedlings are free from competition from other individuals for space and nutrients is of a shorter duration and occurs over a greater area in the devastation zone of Mount St. Helens than at the Nooksack Canyon. The reason for this difference is obviously because water, the major component of glaciers, can exist as a liquid, solid, and gas on the surface of the earth, whereas volcanic deposits exist as gases and liquids only below the surface of the earth. As a result of this physical phenomenon, glaciers advance slowly as solids that later melt and evaporate (or retreat), whereas debris avalanches, pyroclastic flows, and blast surges advance rapidly as liquified or gaslike solids that quickly solidify during deposition. Volcanic disturbances such as pyroclastic flows and mudflows, however, are in a sense similar to glaciers in that a series of smaller episodes may occur subsequent to a catastrophic event. For example, following the May 18 eruption the pyroclastic area north of the Mount St. Helens crater was subjected to a series of episodes of diminishing size resulting in a chronosequence of disturbances. Similarly, the March 19, 1982, eruption of Mount St. Helens produced two mudflows (one into Spirit Lake and the other down the North Fork Toutle) that were smaller than the May 18, 1980, mudflows in the same river basin; therefore, chronosequences of major disturbances do exist, and these result in younger vegetation occurring closer to the volcano. Although the lower Nooksack glacier has been retreating since at least the year 1800, there have been several smaller re-advances (e.g., 1820 and 1970, Fig. 4.3) (Oliver et al. 1979).

Since a glacial recession is temporally extended, plant communities such as those found in the Nooksack Canyon might exhibit a variation in species resulting from changes in the prevailing climate. Plants that invade after or survive a relatively short volcanic cataclysm are more likely to experience a similar climate at similar elevations, exposures, and slopes. Microclimatic effects near glaciers are different from those near volcanoes. Plants that become established during the noninteractive phase of glacial succession have the least competition. The microclimate in these sites, however, is harsh because of cold air drainage from the glacier, and plant species must be able to survive cold as well as low nutrient regimes. Many species characteristic of subalpine communities grow near the receding Carbon, Emmons, and Nooksack montane glaciers of the western Cascades, all located at approximately 1000 m elevation. Nevertheless, the influence of montane glaciers on areas directly below the ice is much less than that demonstrated with continental or large-scale glaciation (Oliver et al. 1985). Following volcanic eruptions, the opposite condition occurs because the area closest to the source of the perturbation (the crater and its associated conduit) is hotter than other sites of equal altitude and aspect. Fumaroles and hot pools are heated by subterranean sources of heat, and it is around these sites with moderate temperatures and moisture that the first recruits begin to grow.

On mudflows and in blowdown areas, microclimate is most variable around small rills or depressions. Here, slower-moving air and more moisture result in less evapotranspiration than from more exposed sites (Gates 1977, Harper 1977). Microclimate is also influenced by soil structure, which varies as a function of the nature and degree of a particular cataclysm. Rocks can serve as thermophiles and alter wind and moisture enough to permit seedling establishment. For example, topographic relief on the Kautz mudflow has affected the revegetation trends in areas that lacked standing dead trees and snags. Here, the lack of microclimate variability resulted in early recruitment of slow-growing tolerant mosses such as *Rhacomitrium canescens* and *R. lanuginosum* and lichens such as *Cladina* and *Stereocaulon* spp. These individuals are able to establish on barren mudflow surfaces and do not have to compete for sunlight with taller plants that may have become established if safe sites were available. Areas on a mudflow older than the 1947 Kautz mudflow were dominated by these initial recruits for many years (fig. 10 of Nelson 1958), and the same situation is found today on the 1947 Kautz mudflow.

SIMILAR EFFECTS OF GLACIATION AND VOLCANISM ON PLANT SUCCESSION

Despite the many differences and complex interrelationships of factors affecting plant succession among the habitats considered, several general conclusions regarding similarities can be made (Table 4.12). These similarities are associated with structure of the substrates, soil chemistry, community structure, secondary succession, and successional models appropriate for each perturbation.

Physical Structure of Substrate

The physical structure of debris slide and pyroclastic material is similar to glacial substrates. Because of the relative dry nature of debris slides, the topography of such deposits is irregular, with flats and mounds that are similar to undulating glacial deposits. A debris avalanche pushes to the side material called lateral terraces and levees. The advancing slide also pushes trees and soil in front of itself, resulting in a mass deposition of vegetation designated the terminal unit (Voight et al. 1981). Pyroclastic flows travel rapidly and deposit levees at the boundary of the flow. The size of these levees varies and is perhaps a reflection of their speed. The lateral terraces of debris slides and the levees of pyroclastic flows are similar to lateral and terminal moraines deposited by glacial advances. For the debris slide of May 18, 1980, the terminal unit is similar to the terminus of an advancing glacier.

Soil Chemistry

Although initial pH values are variable after different disturbances, all soils increase in acidity through time. Ugolini (1968) found initial values of pH at Glacier Bay, Alaska, to be slightly basic, but these declined to values of 4 and 5 through time. On the Kautz mudflow, values were initially near neutral and have dropped at least two orders of magnitude during the last 30 yr (Rigney et al. 1982). Around Mount St. Helens, some soils (debris) were either initially acidic or were rapidly influenced by leaching. Others were originally neutral and decreased in pH by 1 to 4 orders of magnitude within a year, whereas some soils were still neutral, as was the Kautz material for several years.

Nitrogen and phosphorus contents are generally low following both types of catastrophes. Nitrogen content in the Mount St. Helens substrates is higher than in the glacial substrates, yet it is still lower than most soils. Although evidence is limited, cation exchange capacities generally appear to increase with time after disturbances (Ugolini 1968, Ballard 1963, Rigney et al. 1982).

Community Structure

Many of the same species (*Alnus sinuata*, *A. rubra*, *Abies amabilis*, *Tsuga heterophylla*, *T. mertensiana*, *Lupinus latifolius*, *L. lepidus*, and *Epilobium angustifolium*) are found after both glacial and volcanic disturbances in the Pacific Northwest, primarily because of the moderating effect of the maritime climate. These species can play identical successional roles on the two disturbance types. For instance, *Rhacomitrium* spp. are found on mudflows and may be used as indicator species of glacial moraines. In contrast, other species can fulfill different successional roles following glacial or volcanic activity. *Pinus contorta* var. *contorta* grows as mixed stands on river bars and is an ephemeral colonizer of new glacial substrates but is virtually the sole occupant of the overstory of forests on old pyroclastic flows above Goat Marsh (Franklin et al. 1972). *Tsuga mertensiana* is transitional in the Nooksack Canyon where it colonizes sites near receding glaciers and ultimately is replaced by slower-growing, shade-tolerant *A. amabilis*. *Tsuga mertensiana* can also be self-perpetuating at sites where it dominates in the southern Cascades and is climactic in the sense of Clements (1916, 1936). Species that are found in both disturbance types generally have broad niches, are able to tolerate a variety of disturbances, and rapidly immigrate into areas opened by perturbations.

Secondary Disturbances

Ongoing or secondary disturbances have an important influence on communities following volcanic and glacial cataclysms. In Nooksack Canyon, the most common forms of perturbation are snow avalanches that cascade off 1000-m bluffs into the canyon floor. The effects of snow slides on successional trends depend on the size, frequency, and nature of the slide (slab versus dry). The structure and composition of the plant communities struck by snow slides influence the nature of succeeding communities. Many plants (such as *Abies amabilis* and *Alnus sinuata*) possess traits that aid them in surviving and responding to disturbances such as snow slides and wind throws. If a slide occurs before a forest has been established, a shrub community may then be perpetuated. When a slide strikes a forest with a closed canopy, a shrub or herbaceous community may again result. Should a slide occur at the time of understory reinitiation, forest structure may be quickly reestablished by regeneration. Shade-tolerant species such as *Abies amabilis* may be released from suppression by the overstory, and the physical structure of the forest may quickly reemerge (Oliver 1981); thus these communities are resilient to local disturbances. If no subsequent snow slides occur, then an old-growth mix of *Tsuga*, *Pseudotsuga*, and *Abies* results. Around Mount St. Helens, in areas such as the upper Clearwater and Castle watersheds, small *A. amabilis* were suppressed by the overstory. On May 18, 1980, these trees were protected by snow and by their location on a ridge facing away from the steam and ash blast, and they were released from suppression by the blowdown of the overstory.

Erosion is a common disturbance after glacial recession at the Nooksack Canyon and after the lateral blast of Mount St. Helens. The effect of erosion on various substrates, including those produced by the same cataclysm, may be quite different. For example, erosion usually creates havoc after glacial recession by removing fine till, but in the case of Mount St. Helens, it can benefit some plants by releasing buried survivors or by creating safe sites (Harper 1977). As discussed previously, small rills provide safe sites for some species to germinate and persist. When erosion is negligible (as occurred on parts of the Kautz mudflow), lichen and moss ground cover can become established and inhibit recruitment of vascular plants including trees. Moderate erosion provides heterogeneity in substrates and uncovers surviving plants, both of which increase the number of species allowed to coexist. Too little erosion limits seedling success; too much erosion disrupts plant root systems.

In summary, disturbances affect successional processes because individual plants respond to disturbances in various ways (Dyrness 1973, Heinselmann 1973). Localized disturbances in the Mount St. Helens area and at Nooksack Canyon appear to have effects on diversity similar to those reported by other workers (Dayton 1971, Quinn 1979, Platt 1975, Paine and Levin 1981), but for different reasons. In general, moderate perturbations at moderate rates increase diversity, whereas low or high rates and/or magnitudes of disturbance decrease diversity.

Successional Models

Differences in opinion regarding succession have been prevalent throughout this century (Clements 1916, Gleason 1926, Egler 1954, Drury and Nisbet 1973). In contrast to the relay floristics or facilitation model developed on the basis of structural similarities of communities undergoing rapid compositional changes (Cowles 1899, Clements 1916, Dansereau 1954, Olson 1958, Odum 1969), the tolerance model predicts that the sequence of species is solely determined by the characteristics of the life histories of the species. Tolerance to environmental factors fosters net growth of individuals of a given species as they coexist with pioneer species, which later die or decrease in importance because of inferior competitive ability. The inhibition model proposes that early colonists inhibit the establishment of later immigrants. Subsequent colonists are only successful when the preestablished and dominant residents are damaged or killed (Connell and Slatyer 1977). The cyclic model is based on adaptation by certain individuals to recurrent disturbances and the effect of disturbance frequency on community structure (Margalef 1962, Loucks 1970, Dayton 1971, Platt 1975, Horn 1976, Whittaker and Levin 1977).

The characteristics and growth patterns of trees found on volcanic and glacial substrates of the Pacific Northwest provide evidence for the tolerance, inhibition and cyclic models of succession. Native trees appear to be able to survive on mudflow substrates with no prior conditioning by early successional plants. Also, high germination rates in full sunlight, light-weight seeds dispersed long distances, and quick establishment resulting from fast growth can all give species (e.g., *Alnus rubra, Pseudotsuga menziesii,* and *Salix* spp.) a competitive edge over plants with heavier, fewer, and less viable seeds and seedlings with low initial growth rates (e.g., *Abies amabilis* and *A. procera*).

Although trees can grow better when associated with or preceded by forbs, herbs, or shrubs, the presence of numerous saplings on glacial substrates less than 22 yr old, the rapid appearance of conifers on mudflows (e.g., on the 1947 Kautz mudflow) and the debris slide at Mount St. Helens, and the successful germination and growth of trees on North Fork Toutle mudflow material (Table 4.7) suggest that some species characteristically found in later stages of succession can regenerate soon after the disturbance but are simply not noticed due to their small stature as seedlings or do not occur due to the large distance from their seed source. Subsequently, these seedlings may be outcompeted or prevented from becoming established by other species. The presence of *Abies* spp. in later stages of succession may reflect their tolerance to changed conditions such as increasing shade and to the initial inhibition of growth caused by more rapidly established species. It may not reflect the prior existence by species that make the environment more favorable for them. Although conifer phenotypic plasticity has long been known, many contemporary successional studies have ignored this possibility in their consideration of successional mechanisms.

The mortality of trees in the frozen parts of the moisture stress troughs suggests that the limiting factor for success in some habitats is not due solely to the common features of a habitat, but that it is also related to extreme events affecting the roots. Water freezing around roots can be a limiting factor for some species on mudflows as well as in wet habitats below glaciers. In subalpine habitats of the Pacific Northwest, conifer establishment is often limited to mound tops where drainage is relatively rapid. Brink (1959) described such a situation at Black Tusk, British Columbia, and noted that recruitment of conifers was independent of time since glacial recession but did appear to be a function of early exposure from snow such as occurs on mound tops. A debris slide that covered Paradise on the south slope of Mount Rainier still demonstrates tree clumps localized to mound tops. Both the hummocky surface of the debris slide of Mount St. Helens and the morainal mounds of the Nooksack Canyon have topography of high, well-drained rises and slumps with poor drainage. Conifer establishment on mound tops is expected to be more common than in wetter depressions, possibly because of less frequent freezing and thawing.

CONCLUSIONS

Areas in the Pacific Northwest subject to mudflows in the past are currently supporting stands of *Pseudotsuga menziesii*, *Tsuga* spp., *Abies* spp., *Pinus contorta* var. *contorta*, *Alnus rubra*, *Salix* spp., *Populus trichocarpa*, and other species (Frehner 1957, Rigney et al. 1982, Beardsley and Cannon 1930, Franklin et al. 1972, Hemstrom 1979, Heath 1966). These observations, in conjunction with greenhouse results and detailed field studies, indicate that native trees will invade and survive on the recent substrates deposited by the May 18, 1980, eruption of

Mount St. Helens, but with variable rates of success. Initial colonizers will preempt resources so that species with heavier seeds, lower germination rates, and slower initial growth rates may not be able to compete at first. Since *Alnus rubra* and *Pseudotsuga menziesii* produce large seedlings from small seeds and have light-weight seeds, these species would be expected to be early colonists on mudflows and to be better competitors. Other tree species that might be expected to successfully invade the debris slide and mudflows because of their similar life history characteristics include *Populus trichocarpa, Salix sitchensis, S. scouleriana,* and *Alnus sinuata.* Because of the secondary nature of succession in the blowdown area, seedling establishment may be a function of the number of seeds provided by survivors close to the "safe sites." Many species that survived the May 18, 1980, lateral blast of Mount St. Helens have seed and seedling traits favorable for rapid recruitment and establishment in disturbed areas; thus they may have a head start over species that were extirpated or had few individuals to survive. As with other eruptions (Smathers and Mueller-Dombois 1974, Manko 1974), seeds and seedlings had no significant effect on community structure during the first growing season following the catastrophic eruption of May 18, 1980. Yet seedlings have become established in all disturbance habitats and will play an increasingly significant role in the future. Nevertheless, vegetative reproduction will remain important, and *Rubus* spp., *Epilobium* spp., *Anaphalis margaritacea,* and *Lupinus latifolius* are expanding their canopy cover via shoot production from rapidly expanding subterranean organs.

The magnitude, type, and frequency of ongoing disturbances affect the distribution, characteristics, and number of species that coexist. Inhibition and tolerance models of succession are supported by observations of prolonged residency by mosses and lichens on sites suitable for vascular plants and advanced regeneration of *Abies amabilis.* Such interpretations may be more consistent with concepts of natural selection than are models based solely on facilitation, because the initial colonizers actually maintain a site that is more suitable for themselves than for other species. Thus these individuals may live longer and produce more offspring.

Plant successions following volcanic eruptions and glacial recessions in the Pacific Northwest have some features in common. Similarities in species composition following such cataclysms may be due to the prevalence of the maritime climate and to the ability of the plants to respond to similar types of localized disturbances. Ongoing perturbations and soil characteristics influence the rate of succession and the particular species found at each successional stage. There are major differences between the recovery processes following volcanic activity and those following ice retreat. The presence of residual plants or a chronosequence, as well as the type of disturbance, affect the microclimate and diversity of seeds. The overall diversity of deposits found after the Mount St. Helens eruption is greater than that of montane glacial deposits, and some soil changes are occurring faster than had been expected. All of these factors can influence species composition changes through time and create a mosaic of seres. Although comparisons of glacial succession with volcanic succession are useful, time frames and the roles played by various species are not necessarily the same. Information concerning the natural vegetation of Washington and Oregon (Franklin and Dyrness 1973) and the unique conditions created by the Mount St. Helens eruption may provide another independent means of predicting plant succession and of testing current models of forest succession (Shugart and West 1980, Hemstrom and Adams 1982).

ACKNOWLEDGMENTS

We express gratitude to J. F. Franklin, A. R. Kruckeberg, and C. D. Oliver for advice and support. The Geophysics Program, University of Washington, has been most helpful by allowing access to computer resources and by providing accurate information on the eruptive status of Mount St. Helens. F. Swanson, D. Jamison, B. Wilson, E. Small, the U.S. Forest Service, and the Department of Natural Resources Division of Land Management provided logistical support, resources, and ideas. J. Creager, C. Grier, J. Nishitani, R. T. Paine, and R. B. Walker provided constructive criticisms and lab space for soil analysis. We are especially grateful to Warren Tanaka for locating some species in 1984, especially *Collomia debilis*. KING and KIRO television stations in Seattle, the *Everett Herald News*, the *New York Times*, the *Portland Oregonian*, and KATU television station in Portland, Oregon, gave helicopter support and should be applauded for covering scientific research projects. We thank H. Williams (*Seattle Times*) and R. Zazoski for suggesting the topic of this paper. The Mathematics Department of the University of Washington provided computer access. Funding and volunteer support came from Earthwatch (Boston) and The School for Field Studies (Cambridge). We thank the Weyerhauser Co. for allowing access to private land. Some of the work was funded through National Park Service Contract #CX-9000-6-0148 and National Science Foundation Grants DEB-80-04445, DEB-73-03258, and DEB-80-21460.

LITERATURE CITED

ADAMS, V. D., and A. B. ADAMS. 1982. Initial recovery of the vegetation on Mount St. Helens. In *Mount St. Helens: One Year Later Symposium* (S. A. C. Keller, ed.). Eastern Wash. State Univ. Press.

BAKER, H. G. 1974. The evolution of weeds. *Ann. Rev. Ecol. Sys. Ann. Rev.* 5:1–24.

BALLARD, T. M. 1963. Some chemical changes accompanying soil development of the Kautz Creek mudflows, Mount Rainier National Park. M.S. Thesis, Univ. of Washington, Seattle, 75 pp.

BARDO, K. S. 1980. Soil development on neoglacial deposits in the Nooksack Cirque Area. M.S. Thesis, Univ. of Washington, Seattle, 143 pp.

BEARDSLEY, G. F., and W. A. Cannon. 1930. Note on the effects of a mudflow at Mount Shasta. *Ecology* 11:326–336.

BRINK, V. C. 1959. A directional change in the subalpine forest–heath ecotone in Geribaldi Park, British Columbia. *Ecology* 40:10–16.

CARKIN, R. E., J. F. FRANKLIN, J. BOOTH, and C. E. SMITH. 1978. Seeding habits of upper-slope tree species. IV. Seed flight of noble fir and Pacific silver fir. *USDA Forest Service Res. Note PNW* 312:1–10.

CLEMENTS, F. E. 1916. Plant succession: An analysis of the development of vegetation. *Carnegie Inst. Wash. Publ.* 242, 512 pp.

———. 1936. Nature and structure of the climax. *J. Ecol.* 24:252–284.

CLEVELAND, W. S. 1979. Robust locally weighted regression and smoothing scatter plots. *J. Am. Stat. Assoc.* 74(368):829–836.

CLEVELAND, W. S., and I. J. TERPENNING. 1982. Graphical methods for seasonal adjustment. *J. Am. Stat. Assoc.* 77(377):52–62.

COLWELL, W. E. 1943. A biological method for determining the relative boron contents of soils. *Soil Sci.* 56:71–94.

CONNELL, J. H., and R. O. SLATYER. 1977. Mechanisms of succession in natural communities and their role in community stability and organization. *Amer. Nat.* 111:1119–1144.

COWLES, H. C. 1899. The ecological relations of the vegetation on the sand dunes of Lake Michigan. *Bot. Gaz.* 27:95–117, 167–202, 281–308, and 361–391.

CRANDELL, D. R., and H. H. WALDRON. 1956. A recent volcanic mudflow of exceptional dimensions from Mount Rainier, Washington. *Amer. J. Sci.* 254:349–362.

CREAGER, J. S., P. A. McMANUS, and E. F. COLLIAS. 1962. Electronic data processing in sedimentary analysis. *J. Sed. Petrol.* 32:833–839.

DANSEREAU, P. 1954. Studies on Central Baffin Vegetation. I. Bray Island. *Vegetatio* 5:329–339.

DAYTON, P. K. 1971. Competition, disturbance, and community organization: The provision and subsequent utilization of space in a rocky intertidal community. *Ecol. Monogr.* 41:351–389.

DECKER, R., and B. DECKER. 1981. The eruption of Mount St. Helens. *Sci. Amer.* 244:68–80.

DENTON, C. H., and S. C. PORTER. 1970. Neoglaciation. *Sci. Amer.* 222:100–110.

DRURY, W. H., and I. C. T. NISBET. 1973. Succession. *J. Arnold Arb.* 54:331–368.

DYRNESS, C. T. 1973. Early stages of plant succession following logging and burning in the western Cascades of Oregon. *Ecology* 51:57–69.

EGLER, R. E. 1954. Vegetation science concepts. I. Initial floristic composition, a factor in old-field vegetation development. *Vegetatio* 4:412–417.

FRANKLIN, J. F., F. C. HALL, C. T. DYRNESS, and C. MASER. 1972. *Federal Research Natural Areas in Oregon and Washington, A Guidebook for Scientists and Educators.* USDA Forest Service PNW Forest and Range Exp. Station, Portland, Oregon.

FRANKLIN, J. F., and C. T. DYRNESS. 1973. *Natural Vegetation of Oregon and Washington.* USDA Forest Service, Gen. Tech. Report PNW-8. 417 pp.

FREHNER, H. F. 1957. Development of soil and vegetation on Kautz Creek flood deposit in Mount Rainier National Park. M.S. Thesis, Univ. of Washington, Seattle, 83 pp.

GATES, D. M. 1977. Transpiration and leaf temperature. Manuscript, Center for Quantitative Science, Univ. of Washington, Seattle.

GLEASON, H. A. 1926. The individualistic concept of the plant association. *Bull. Torrey Bot. Club* 53:7–26.

GOLDSTEIN, R. A., and D. R. GRIGAL. 1972. *Computer Programs for the Ordination and Classification of Ecosystems.* Oak Ridge Nat. Lab., Environ. Sci. Div. Pub. 417, 125 pp.

GOLDTHWAIT, R. P., F. LOEWE, F. C. UGOLINI, H. F. DECKER, D. M. DELONG, M. B. TRAUTMAN, E. E. GOOD, T. R. MERRELL, and E. O. RUDOLPH. 1966. *Soil Development and Ecological Succession on Deglaciated Area of Muir Inlet, S.E. Alaska.* Ohio State Univ. Research Foundation, Columbus, Ohio, 166 pp.

GRIGGS, R. F. 1917. The valley of ten thousand smokes. *Nat. Geog. Mag.* 21:13–68.

GROSVALD, M. G., and A. F. GLAZOVSKII. 1981. The interaction of glaciation and volcanism and its manifestation in the behavior and morphology of glaciers. In *The Interaction between Volcanism and Glaciation* (seminar), Akademia NAUK, Petropablovsk–Kamchatka.

HARPER, J. L. 1977. *Population Biology of Plants.* Academic Press, New York, 892 pp.

HEATH, J. P. 1966. Primary conifer succession: Lassen Volcanic National Park. *Ecology* 48:270–275.

HEINSELMANN, M. L. 1973. Fire in the virgin forests of the Boundary Water Canoe Area, Minnesota. *J. Quat. Res.* 3:329–382.

HEMSTROM, M. L. 1979. A recent disturbance history of forest ecosystems at Mount Rainier National Park, Washington. Ph.D. Dissertation, Oregon State Univ., Corvallis, 67 pp.

HEMSTROM, M. L., and V. D. ADAMS. 1982. Modeling long-term forest succession in the Pacific Northwest. In *Proc. Symp. Forest Succession and Stand Development Research* (J. Means, ed.), 14–23, Forest Research Laboratory, Oregon State Univ., Corvallis.

HITCHCOCK, C. L., A. CRONQUIST, M. OWNBEY, and J. W. THOMPSON. 1973. *Vascular Plants of the Pacific Northwest.* Univ. of Washington Press, Seattle.

HORN, H. S. 1976. Succession. In *Theoretical Ecology* (R. M. May, ed.), 187–190, W. B. Saunders, Philadelphia.

HUTCHINSON, G. E. 1951. Copepodology for the ornithologist. *Ecology* 32:571–577.

———. 1959. Homage to Santa Rosalia, or why are there so many kinds of animals. *Amer. Nat.* 93:145–149.

JENNY, H., J. VLAMIS, and W. E. MARTIN. 1950. Greenhouse assay of fertility of California soils. *Hilgardia* 20:1–8.

KACHURA, N. N. 1974. Influence of the aboveground cutting on the growth, development and regeneration of *Filipendula camtschatica* (Pall.) Maxim. *Komarov Readings (Vladivostok)* 22:61–77.

KRUCKEBERG, A. R. 1951. Intraspecific variability in the response of certain native plant species to serpentine soil. *Amer. J. Bot.* 38:408–419.

———. 1954. The ecology of serpentine soils: A symposium. III. Plant species in relation to serpentine soils. *Ecology* 35:267–274.

———. 1967. Ecotypic response to ultramafic soils by some plant species of northwestern United States. *Brittonia* 19:133–151.

———. 1980. Mount St. Helens foray. *Douglasia.* 3:9–10.

LAWRENCE, D. B. 1954. Diagrammatic history of the northeast slope of Mount St. Helens, Washington. *Mazama* 36:41–44.

LIPMAN, P. W., and D. R. MULLINEAUX (eds.). 1981. In *The 1980 Eruptions of Mount St. Helens, Washington.* U.S. Geol. Surv. Prof. Paper 1250, U.S. Govt. Printing Office, Washington, D.C., 844 pp.

LOUCKS, O. L. 1970. Evolution of diversity, efficiency and community stability. *Amer. Zool.* 10:17–25.

MACARTHUR, R. H., and E. O. WILSON. 1967. *The Theory of Island Biogeography.* Princeton Univ. Press, Princeton, New Jersey. 203 pp.

MACK, R. N. 1981. Initial effects of ashfall from Mount St. Helens on vegetation in eastern Washington and adjacent Idaho. *Science* 213:537–539.

MANKO, Yu. I. 1974. The effect of contemporary vulcanism on the vegetation of Kamchatkas and the Kurile Islands. *Komarov Lectures (Vladivostok)* 22:5−31.

——. 1975. Recent volcanic activity: A factor in vegetation dynamics. In *13th Pacific Sci. Congr.*, I., 203−204 (abstr.). Vancouver, British Columbia, Canada.

MARGALEF, R. 1962. Successions of populations. *Adv. Front. Plant Sci.* 2:137−188.

MEANS, J. E., W. A. McKEE, W. H. MOIR, and J. F. FRANKLIN. 1982. Natural revegetation of the northeastern portion of the devastated area. In *Mount St. Helens: One Year Later Symposium* (S. A. C. Keller, ed.), Eastern Wash. State Univ. Press, pp. 93−103.

MUELLER-DOMBOIS, D., and H. ELLENBERG. 1974. Aims and Methods of Vegetation Ecology. John Wiley, New York. 547 pp.

NELSON, J. W. 1958. Soil properties related to vegetation and time on the Kautz Creek flood area, Mount Rainier, Washington. M.S. Thesis, Univ. of Washington, Seattle, 56 pp.

ODUM, E. P. 1969. The strategy of ecosystem development. *Science* 164:262−270.

OLIVER, C. D. 1981. Forest development in North America following major disturbances. *Forest Ecol. Manage.* 3:153−168.

OLIVER, C. D., and A. B. ADAMS. 1979. Vegetation dynamics. In *Natural History of the Nooksack Cirque* (C. D. Oliver et al., eds.), 42−96, Natl. Park Service Report CX-9000-6-0418.

OLIVER, C. D., R. WEISBROD, R. ZAZOSKI, A. B. ADAMS, K. BARDO, and J. DRAGAVON (eds.). 1979. *Natural History of the Nooksack Cirque*. Nat. Park Service Report CX-9000-6-0418, 144 pp.

OLIVER, C. D., A. B. ADAMS, and R. J. ZAZOSKI. 1985. Disturbance patterns and forest development in a recently deglaciated valley in the northwestern Cascade Range of Washington, U.S.A. *Can. J. For. Res.* 15(1):221−232.

OLSON, J. S. 1958. Rates of succession and soil changes on southern Lake Michigan sand dunes. *Bot. Gaz.* 119:125−170.

ORLOCI, L. 1969. Information analysis of structure in biological collections. *Nature* 223:483−484.

PAINE, R. T., and S. A. LEVIN. 1981. Intertidal landscapes: Disturbance and the dynamics of pattern. *Ecol. Monogr.* 51:145−178.

PARKINSON, J. A., and S. E. ALLEN. 1975. A wet oxidation procedure suitable for the determination of nitrogen and mineral nutrients in biological material. *Common Soil Sci. Plant Anal.* 6:1−11.

PEET, R. K. 1974. The measurement of species diversity. *Ann. Rev. Syst. Ecol.* 5:285−308.

PICKETT, A. T. A., and F. A. BAZZAZ. 1976. Divergence of two co-occurring successional annuals on a soil moisture gradient. *Ecology* 57:169−176.

PLATT, W. J. 1975. The colonization and formation of equilibrium plant species associations on badger disturbances in a tall-grass prairie. *Ecol. Monogr.* 45:285−305.

PORTER, S. C. 1981. Recent glacier variations and volcanic eruptions. *Nature* 291:139−142.

PORTER, S. C., and C. H. DENTON. 1967. Chronology of neoglaciation in North America Cordillera. *Amer. J. Sci.* 265:177—210.

QUINN, J. F. 1979. Disturbance, predation and diversity in the rocky intertidal zone. Ph.D. Dissertation, Univ. of Washington, Seattle, 225 pp.

RAUNKIAER, C. 1934. *The Life Forms of Plants and Statistical Plant Geography*. Clarendon Press, Oxford, 632 pp.

REINERS, W. A., I. A. WORLEY, and D. B. LAWRENCE. 1971. Plant diversity in a chronosequence at Glacier Bay, Alaska. *Ecology* 52:55−69.

RICHARDS, L. A. 1954. *Diagnosis and Improvement of Saline and Alkali Soils. Agricultural Handbook 60*, U.S. Govt. Printing Office, Washington, D.C., 100 pp.

——. 1965. Physical condition of water in soil. In *Methods of Soil Analysis, Part I. Physical and Mineralogical Properties, Including Statistics of Measurement and Sampling*. (C. A. Black, D. D. Evans, J. L. White, L. E. Ensminger, eds.), 128−152, Publisher, Inc., Madison, Wisconsin.

RIGNEY, L. P., M. E. KRASNEY, and P. M. FRENZEN. 1982. Patterns of vegetation and soil development on the Jautz mudflow, Mount Rainier National Park, Washington (abstr.), In *54th Northwest Sci. Meeting*, p. 34, Corvallis, Oregon.

Rosenfeld, C. L. 1980. Observations on the Mount St. Helens eruption. *Amer. Sci.* 68:494–509.

Saint John, H. 1976. The flora of Mount St. Helens, Washington. *The Mountaineer* 70:65–77.

Sharitz, R. R. 1973. Population dynamics of two competing annual species. *Ecology* 54:723–740.

Shugart, H., and D. West. 1980. Forest succession models. *Bioscience* 30:308–313.

Smathers, G. A., and D. Mueller-Dombois. 1974. Invasion and recovery of vegetation after volcanic eruption in Hawaii. *Nat. Park Service Sci. Monogr.* 5(38):129.

Stork, A. 1963. Plant immigration in front of retreating glaciers. *Geografiska Annaler* 45:1–22.

Trappe, J. M., J. F. Franklin, R. F. Tarrant, and G. M. Hansen (eds.). 1968. *Biology of Alder.* Proc. Symp. 40th Northwest Sci. Assoc., PNW Forest and Range Exp. Stn. USDA Forest Service, Portland, Oregon.

Turner, N. L., and J. Y. Parlange. 1970. Analysis of operation and calibration of a ventilated diffusion porometer. *Plant Physiology* 46:175–177.

Ugolini, F. C. 1968. Soil development and alder invasion in a recently deglaciated area of Glacier Bay, Alaska. In *Biology of Alder* (J. M. Trappe, J. F. Franklin, G. M. Hansen, and R. F. Tarrant, eds.), 115–140, Proc. Symp. 40th Northwest Sci. Assoc., PNW Forest and Range Exp. Sta. USDA Forest Service, Portland, Oregon.

Ugolini, F. C., and R. J. Zazoski. 1979. Soils derived from tephra. In *Volcanic Activity and Human Ecology* (P. D. Sheets and D. K. Grayson, eds.), pp. 83–124, Academic Press, New York.

United States Weather Bureau. 1980. Climatological Data Annual Summary, Washington 1979. Ashville, N.C., National Climatic Center, 19 pp.

Viereck, L. A. 1966. Plant succession and soil development on gravel outwash on Muldrow Glacier, Alaska. *Ecol. Monogr.* 36:181–199.

Virogradov, V. H. 1981. Volcanism and glaciation: Status of problems and prospects for research. In *The Interaction between Volcanism and Glaciation* (seminar), Akademia NAUK, Petropablovsk–Kamchatka.

Voight, B., H. Glicken, R. J. Janda, and P. M. Douglass. 1981. Catastrophic rockslide avalanche of May 18. In *The 1980 Eruptions of Mount St. Helens, Washington*, 347 (P. W. Lipman and D. R. Mullineaux, eds.), U.S. Geol. Surv. Prof. Paper 1250, U.S. Govt. Printing Office, Washington, D.C.

Voroshilov, V. P., and A. N. Sidel'nikov. 1979. Features of vegetation distribution in the area of the solfatara field of the Mendeleev volcano. *Ekologiya* (6):30–35.

Whittaker, R. H., and S. A. Levin. 1977. The role of mosaic phenomena in natural communities. *J. Theor. Popl. Biol.* 12:117–139.

Winner, W. E., and H. A. Mooney. 1980. Responses of Hawaiian plants to volcanic sulfur dioxide: Stomatal behavior and foliar injury. *Science* 210:789–790.

xxxxxxxxxxxxxxxxxxxxxxxxxxxxxxxxxxxxxxxxxxxxxxxxxxxxxxxxxxxxxxxxxxxxxxxxxxx

TABLE 4.1
Weather Information for Sites in Washington

Station	Elevation (m)	Latitude	Longitude	Average annual temperature (°C)	Average annual precipitation (cm)	Average annual snowfall (cm)
Spirit Lake[a]	1063	46°16'N	122°09'W	5.6	225.3	718
Longmire[a]	842	46°45'N	121°49'W	7.3	140.8	—
Nooksack[b]	1000	48°51'N	121°34'W	5.5	250.0	1003

[a]From the U.S. Weather Bureau (1980).
[b]Interpolation using elevation and climate data from stations at Mount Baker Ski Lodge and Glacier, Washington (U.S. Weather Bureau, 1980).

TABLE 4.2
Particle Size Analysis[a]

Deposit and site	Sand (wt %)	Silt (wt %)	Clay (wt %)	Shepard class	Folk–Ward Values		
					Sortedness	Skewness	Kurtosis
Mud							
Confluence of Cowlitz and Toutle[c]	78	19	3	1	5	5	2
	76	20	4	1	5	5	2[b]
2 km below debris flow[c]	72	24	4	2	5	4	2
	74	23	3	2	5	4	3
	75	21	4	2	5	4	3
	73	22	5	2	5	4	2
	73	22	5	2	5	4	2
	75	21	4	2	5	4	2
Upper South Fork Toutle	82	16	2	1	5	4	2
Kautz (1956)[d]	78	18	4	1	—	—	—
Kautz (1980)	84	14	2	1	5	4	2
Debris							
Castle Lake 15 cm below surface[c]	62	33	4	2	5	3	4

Ash						
Timberline 4/80/30	73	27	1	2	4	3	3
Castle Lake						
debris flow surface	99	0.5	0.5	1	3	1	2
Elk Rock	57	40	3	2	4	3	4
Yakima	89	10	1	1	3	5	5
Spud Mountain 8/80	98	2	0	1	3	3	3
Glacial						
Nooksack						
(deposited 1975)[e]	88	8	4	1	—	—	—

[a]Collected summer of 1980, unless indicated otherwise.
[b]Same collection as the preceding sample, but differs by first being treated with hydrogen peroxide.
[c]Samples were taken from material that was used in greenhouse moisture stress troughs and/or Jenny pot tests.
[d]From Nelson 1958.
[e]From Bardo 1980.

TABLE 4.3

CHARACTERISTICS FOR MOUNT ST. HELENS SOILS

Deposit and site	Collection date	Physical properties				
		Rocks 2 mm (%)	pH 1:1 H_2O	EC^a (mmhos/cm)	Saturation (% per 15 atm)	Moisture release curve: $Y = g/ H_2O/g$ soil $= \dot{N}_2$ gas
Mud flows (7/80) Confluence of Cowlitz and Toutle	1/81	—	5.6	0.89	5.9	$r^2 = 0.88$, p 0.001
2 km below debris slide	6/80	—	6.8 0.02[b]	1.96	5.8	—
	7/80	6.0	—	0.55	Not available	—
Debris slide flats Castle Lake 15 cm below surface	7/80	9.5	4.4 0.15[b]	1.57	6.4	$Y = 0.07$ $^{-0.20}$, $N = 22$, $r^2 = 0.88$, p 0.001
Central flats	8/81	17.9	5.0	0.28	5.3	$Y = 0.06$ $^{-0.21}$, $N = 10$, $r^2 = 0.95$, p 0.001

Upper flats						
between Castle and Coldwater lakes	8/81	30.3	5.3	1.83	3.9	— —
	8/81	4.22	4.5	0.47	5.9	
Debris slide mounts						
NW cove						
Spirit Lake	9/81	64.4	5.2	0.10	5.6	— —
Jackson Lake	9/81	63.0	5.1	0.79	4.5	
Ash						
Spud Mountain	8/80	.01	6.4	0.30	57.1	$Y = 0.02\psi^{-0.39}$, $N = 26$, $r^2 = 0.74$, $p < 0.001$ —
Crater	5/81	—	6.5	0.24	—	—
Pyroclastics						
Mount St. Helens north slope	5/81	24.8	5.6	0.15	6.9	$Y = 0.05\psi^{0.33}$, $N = 21$, $r^2 = 0.70$, $p < 0.001$
Sulfatero	5/81	32.5	2.8	6.49	7.9	—
Goat Marsh	4/81	—	5.5	0.14	8.1	—

[a]EC = electrical conductivity.
[b]Mean ± standard deviation from Table 4.4.

TABLE 4.4
SOILS FROM THE NORTH FORK OF THE TOUTLE RIVER MUDFLOWS AND DEBRIS SLIDE, KAUTZ MUDFLOW AND THE NOOKSACK GLACIAL AREA[a]

Deposit and site	Date of collection	pH 1:1 H_2O	Weight loss ignition (%)	Total carbon (ppm)	Nitrogen (NH_3) (ppm)	C/N
				Organic properties of soils		
Mud flows						
Confluence of						
Cowlitz	8/80	5.9	0.71	1050	95	11.11
and Toutle	7/80	6.0	0.36	760	423	1.80
2 km below	8/80	5.8	0.71	2960	100	29.60
debris	9/80	4.7	0.24	690	24	28.16
slide	7/80	6.0	0.64	910	37	24.27
Kautz[b]	1951	6.5	NG[c]	NG	10	NG
Kautz[d]	1980	5.2	NG	NG	160−560	NG

Debris slides							
Castle Lake 15 cm below surface	7/80	4.1	0.31	160	703	0.23	
		4.2					
		4.8					
Ash							
Castle Lake debris slide	7/80	5.8	0.38	1420	76	18.78	
surface	9/80	5.4	0.32	350	203	1.72	
Glacial							
Nooksack (time zero)[e]	1976	4.91	NG	600	8	75.00	

[a]Collected summer of 1980 unless indicated otherwise.
[b]From Frehner 1957. Values are for A horizons.
[c]NG = not given.
[d]From Rigney et al. 1982. Values are for A horizons.
[e]From Bardo 1980.

TABLE 4.5
ANALYSIS OF VARIANCE OF LETTUCE WEIGHTS FROM JENNY POT
TEST OF MOUNT ST. HELENS SOILS

Source of variation	Sum of squares	D.F.	Mean square	F	Significance
Main effects	49.515	8	6.189	55.096	0.001
Soil	18.850	2	9.425	83.901	0.001
Treatments	30.708	6	5.118	45.558	0.001
Two-way inter-actions	16.941	12	1.412	12.567	0.001
Explained	66.456	20	3.323	29.579	0.001
Residual	7.007	63	0.112	—	—
Total	73.534	83	0.886	—	—

TABLE 4.6

Mean Dry Weight of Lettuce Plants (g) with Relative Yields in Parentheses[a]

Material	Distance from crater (km)	Treatment						
		None	$N_2P_4K_{1/2}$	$N_0P_4K_{1/2}$	$N_2P_0K_{1/2}$	$N_2P_4K_0$	$N_2P_8K_{1/2}$	$N_2P_8K_{1/2}$ plus micronutrients
Debris slide at Castle Lake 15 cm below surface	7	.01 (16.67)	0.06 (100.00)	0.06 (100.00)	0.12 (200.00)	0.10 (66.67)	0.22 (366.67)	0.14 (233.33)
Mudflow 2 km below terminus of debris slide	28	0.01 (0.55)	1.83 (100.00)	0.01 (0.55)	0.05 (2.73)	2.20 (12.02)	1.56 (85.25)	2.11 (115.30)
Confluence of Cowlitz and Toutle	50	0.17 (8.85)	1.92 (100.00)	0.30 (15.63)	0.08 (4.17)	2.37 (123.44)	1.88 (97.91)	1.00 (52.08)

[a]Relative yield is the absolute weight of the treatment divided by the weight of the complete treatment $(N_2P_4K_{1/2}) \times 100$.

TABLE 4.7
Seed and Seedling Characteristics of Native Trees

Species	Successional status[a]	Germination On mud (%)	Optimum[b] (%)	Seed weight (g)	Seedling weight Fresh (g)	Dry (g)	Amount above ground (fresh) (%)	Total dry weight seed/seedling ratio
Thuja plicata	Late	42	74	0.0142 (0.0071)[c]	0.0409 (0.0078)	0.0075 (0.0013)	49.07 (11.76)	0.528
Pseudotsuga menziesii	Seral	64	89	0.0896 (0.0393)	0.4626 (0.1386)	0.0792 (0.0204)	34.70 (4.29)	0.883
Pinus contorta	Early seral	66	96	0.0369 (0.0077)	0.1219 (0.0376)	0.0266 (0.0084)	35.75 (10.00)	0.720
Tsuga heterophylla	Late	22	60	0.0200	0.0578 (0.0183)	0.0114 (0.0030)	65.90 (12.28)	0.570
Abies amabilis	Late	16	36	0.4670 (0.1009)	0.3111 (0.0404)	0.0651 (0.0106)	50.33 (5.94)	0.139
Abies procera	Seral	6	27	0.4270	0.3919	0.0720	58.05	0.169
Alnus rubra	Early seral	23	66	0.0095 (0.0053)	0.1170 (0.0617)	0.0342 (0.0146)	33.56 (16.40)	3.600

[a] Successional status is based on Franklin and Dyrness (1973).
[b] Optimum germination was obtained by placing seeds on moist filter paper with adequate light (data from the Dept. of Natural Resources, Washington State).
[c] Standard deviation in parentheses.

TABLE 4.8

LIFE FORMS BY SPECIES (%) FOR VARIOUS SITES AT
MOUNT ST. HELENS, KAUTZ CREEK, AND NOOKSACK CIRQUE

Site	Phaner-ophytes	Chamae-phytes	Hemicryp-tophytes	Geo-phytes	Thero-phytes
Mount St. Helens[a]					
Total area	22.9	4.3	12.9	57.0	2.9
Mudflows	63.7	0.0	0.0	36.3	0.0
Blowdown	19.7	5.4	10.7	60.6	3.6
Singe	27.3	3.0	21.2	48.5	0.0
Kautz Creek[b]	22.3	6.7	15.5	51.1	4.4
Nooksack Cirque[c]					
Communities	24.6	12.3	13.7	49.4	0.0
Eroded till	20.6	14.7	17.6	47.1	0.0
Rubble meadows	19.4	11.3	14.5	54.8	0.0
Morainal mounds	46.1	19.2	11.5	23.2	0.0

[a]Devastation Zone.

[b]Kautz mudflow, 7 yr after the mudflow; data based on species list in Frehner (1957).

[c]Exposed by glacier for < 33 yr.

TABLE 4.9
PERCENTAGE GERMINATED (%G) AND PERCENTAGE ALIVE (%A) IN MAY, 1981, OF 200 SEEDS PLANTED FALL OF 1980 AT FOUR SITES IN THE MOUNT ST. HELENS VICINITY

Site	Populus tricocarpa		Alnus rubra		Abies amabilis		Abies procera		Pseudotsuga menziesii		Pinus contorta		Tsuga heterophylla		Thuja plicata	
	%G	%A	%G	%A	%G	%A	%G	%A	%G	%A	%G	%A	%G	%A	%G	%A
Castle Lake (ash)	0	0	9.0	0.5	0	0	0	0	0	0	0	0	0.5	0.5	1.0	0
Coldwater Lake (ash)																
open	0	0	0	0	0	0	NP[a]		2.0	0	6.0	0	NP		0.5	0
sheltered	1.0	0.5	0	0	0	0	NP		0	0	2.0	1.0	NP		0.5	0.5
Central debris slide	WO[b]		WO		WO		WO		WO		WO		WO		WO	
Debris slide, N.W. cove of Spirit Lake	0	0	0	0	WO		WO		15.5	4.5	WO		WO		WO	
						Number of seedlings from natural colonization										
Debris slide	0		0		2		1		3		0		2		0	

[a]NP = not planted.
[b]WO = washed out.

TABLE 4.10
NUMBER OF SEEDLINGS ON NORTH FORK TOUTLE DEPOSITS

Species in 25 = m square plots	Pyroclastic (2 plots)	Debris slide Castle–Coldwater (31 plots)	Debris slide below Elk Rock (18 plots)
Epilobium angustifolium	—	1000+	6
Salix spp.	4	4	183
Carex spp.	1	—	181
Cirsium arvense	—	1	95
Senecio sylvaticus	—	—	42
Equisetum spp.	2	20	2
Petasites frigidus	—	3	16
Agrostis sp.	—	—	7
Vaccinium membranaceum	—	—	6
Rubus ursinus	1	1	4
Festuca sp.	—	5	—
Populus trichocarpa	—	—	5
Hypochaeris radicata	2	—	2
Calamagrostis sp.	—	—	3
Lupinus latifolius	—	—	3
Tolmiea menziesii	—	—	3
Trifolium microcephalum	—	—	3
Abies amabilis	—	2	—
Cardamine oligosperma	—	—	2
Alnus rubra	—	—	1
Anaphylis margaritacaeae	—	1	—
Digitalis purpureae	—	—	1
Epilobium alpinum	1	—	—
Juncus sp.	—	—	1
Sonchus arvensis	—	—	1
Tsuga heterophylla	—	1	—

TABLE 4.11
DISTURBANCES AND THEIR EFFECTS ON VARIOUS FACTORS OF PLANT SUCCESSION

Factors	Glacial (Nooksack <37 yr old)	Volcanic				
		Blowdown and singe	Mudflow	Debris slide	Pyroclastic flow	Mudflow (Kautz)
A. Structure	Rubble, moraines	Sand at surface	Hetero-geneous	Hetero-geneous	Variable levees	Hetero-geneous
B. Soil chemistry						
Initial pH	Acidic	Neutral	Neutral	Acidic to neutral	Acidic to neutral	Neutral
Conductivity	Not given	<2	0.1–0.2	0.1–2.0	0.15–6.5	Not given
Nitrogen[a]	1	2	3 (most)	3	0	3

	None	Stumps, downed trees	Snags, root wads	Snags, root wads	None	Snags, root wads
C. Community structure and recruitment						
Organic remains	None	Stumps, downed trees	Snags, root wads	Snags, root wads	None	Snags, root wads
Residual plants						
Number of plants	0	3 (most)	2	1	0	1?
Number of species	3	4 (most)	2	1	0	2
Number of seeds	3 (most)	1	2	1	1	3
D. Chronosequence	Yes	None	None	None	None	None
E. Microclimatic effects	Cold close to glacier	Shelter important	Shelter important	Shelter important	Fumaroles	Shelter important

[a]Relative scale.

TABLE 4.12

CHEMICAL, STRUCTURAL, AND FLORISTIC SIMILARITIES OF DEPOSITS AND
TRENDS FOLLOWING GLACIAL RETREATS AND VOLCANIC ERUPTIONS

Similarities of deposits

1. The <2-mm fraction has a large proportion of sand.
2. Disturbances such as debris slides push pieces of forests and soil in front of themselves in a similar way to glacial advances, resulting in piles of organic matter being deposited at the terminus. Structures such as lateral terraces and hummocks left by relatively dry debris slides are similar to lateral moraines and mounds deposited by glaciers.
3. Following glacial retreat and some types of volcanic disturbances (e.g., lava flows, pyroclastic flows, and debris slides), residual survivors are lacking or rare.
4. Nitrogen and phosphorus levels are initially low.

Similarities of trends through time

5. Acidity increases through time, although initial values are variable.
6. Cation exchange capacity increases with time.
7. Species composition and life form spectra are similar.
8. Ongoing disturbances affect successional trends. The mechanisms by which plants survive are similar subsequent to local disturbances accompanying glacial retreats and volcanic eruptions.

APPENDIX 4.A
Nutrient Additions to Jenny Pot Test

Treatment	Chemical addition	Amount (ml)
Control	None	
$N_2P_4K_{1/2}$	$NH_4H_2PO_4$ (10%)	5.17
	KNO_3 (5%)	1.70
	NH_4NO_3 (10%)	2.42
$N_0P_4K_{1/2}$	NaH_2PO_4 (10%)	6.21
	KCL (5%)	1.26
$N_2P_0K_{1/2}$	KNO_3	1.70
	NH_4NO_3	4.23
$N_2P_4K_0$	NH_4NO_3	2.74
	$NH_4H_2PO_4$	5.17
$N_2P_8K_{1/2}$	KNO_3	1.70
	$NH_4H_2PO_4$	10.35
	NH_4NO_3	0.61
$N_2P_8K_{1/2}$ plus micronutrients	KNO_3	1.70
	$NH_4H_2PO_4$	10.35
	NH_4NO_3	0.61
	Na_2SO_4	1.46
	$MgCl_2$	0.67
	A_5	1.33

SPECIES LIST BY DISTURBANCE TYPE SHOWING MEAN PERCENTAGE OF COVER (%C) AND RELATIVE FREQUENCY (%F)[a]

Species	Mount St. Helens (Summer 1980)[b]						1954 Kautz mudflow (%F)[c]		1977 Nooksack Cirque (%F)[d]		
	Mud		Blowdown		Singe		Under snags	In open	Eroded till	Rubble meadows	Morainal mounds
	%C	%F	%C	%F	%C	%F					
Lycopodiaceae											
Lycopodium alpinum L.	0	0	0	0	0	0	0	0	0	0	7
Selaginellaceae											
Selaginella wallacei Hieron.	0	0	0	0	0	0	0	0	0	13	0
Equisetaceae											
Equisetum sp. L.	+[e]	14	+	7	0	0	0	0	0	0	0
Polypodiaceae											
Asplenium viride Huds.	0	0	0	0	0	0	1	0	0	0	0
Athyrium filix-femina (L.) Roth	+	14	+	0	0	0	78	0	0	0	0
Athyrium distentifolium Tausch	0	0	0	0	0	0	0	0	43	0	0
Cryptogramma crispa (L.) R. Br.	0	0	0	0	0	0	0	0	57	100	13
Dryopteris austriaca (Jacq.) Woynar	0	0	0	0	0	0	56	17	0	0	0

Polypodiaceae cont'd

Polystichum munitum (Kaulf.) Presl	0	0	+	4	2.13	5	33	0	14	0	13
Polystichum andersonii Hopkins	0	0	0	0	0	0	67	0	0	0	0
Pteridium aquilinum (L.) Kuhn.	+	14	0.68	17	0.63	13	56	0	0	0	0

Cupressaceae

Chamaecyparis nootkatensis (D. Don) Spach	0	0	0	0	0	0	0	0	14	38	7
Thuja plicata Donn.	0.35	114	0	0	0.13	0.13	78	0	0	0	7

Pinaceae

Abies amabilis (Dougl.) Forbes	0	0	0	0	0	0	11	17	0	13	73
Abies grandis (Dougl.) Forbes	0	0	0	0	0	0	11	0	0	0	0
Abies lasiocarpa (Hook.) Nutt.	0	0	0	0	0	0	0	0	0	13	0
Abies procera Rehder	0	0	0	0	0.13	13	0	0	0	0	0
Pinus contorta Dougl.	0	0	0	0	0	0	0	0	0	0	13
Pinus monticola Dougl.	0	0	0	0	0	0	33	0	0	0	0
Pseudotsuga menziesii (Mirbel) Franco	2.86	28	+	4	0.13	13	89	50	0	0	7

APPENDIX 4.B
SPECIES LIST BY DISTURBANCE TYPE SHOWING MEAN PERCENTAGE OF COVER (%C) AND RELATIVE FREQUENCY (%F)[a] (continued)

Species	Mount St. Helens (Summer 1980)[b]						1954 Kautz mudflow (%F)[c]		1977 Nooksack Cirque (%F)[d]		
	Mud		Blowdown		Singe		Under snags	In open	Eroded till	Rubble meadows	Morainal mounds
	%C	%F	%C	%F	%C	%F					
Salicaceae											
Populus trichocarpa T. and G.[f]	7.29	86	0.09	9	1.00	100	0	0	0	0	0
Salix scouleriana Barratt	0	0	0	0	0	0	0	0	0	13	7
Salix sitchensis Sanson	0	0	0	0	0.63	13	89	17	0	0	0
Betulaceae											
Alnus rubra Bong.	2.14	14	0	0	0	0	44	0	0	0	0
Alnus sinuata (Regel) Rydb.	0	0	+	4	0	0	0	0	29	50	87
Polygonaceae											
Polygonum kelloggii Greene	0	0	0	0	0	0	0	0	0	13	0
Portulacaceae											
Montia parvifolia (Moc.) Greene	0	0	0	0	0	0	0	0	0	63	0

appendix 4.B continued

Portulacaceae cont'd											
Montia perfoliata (Donn) Howell	0	0	0.17	9	0	0	0	0	0	0	0
Montia siberica (L.) Howell	0	0	0	0	0	0.25	13	0	0	0	13
Caryophyllaceae											
Cerastium sp.	0	0	0	0	0	0	0	1	0	0	0
Ranunculaceae											
Ranunculus repens L.	0	0	0	0	0	0.38	25	0	0	0	0
Aquilegia formosa Fisch.	0	0	0	0	0	0	0	0	0	0	13
Berberidaceae											
Achlys triphylla (Smith) DC.	0	0	+	4	0	0	0	0	0	0	0
Vancouveria hexandra (Hook) Morr. and Dec	0	0	0.22	9	0	0	0	0	0	0	0
Berberis sp.	0	0	0	0	0	0	0	3	0	0	0
Fumariaceae											
Dicentra formosa (Andr.) Walp.	0	0	0.22	9	0	0	0	0	0	0	25

SPECIES LIST BY DISTURBANCE TYPE SHOWING MEAN PERCENTAGE OF COVER (%C)
AND RELATIVE FREQUENCY (%F)[a] (continued)

| Species | Mount St. Helens (Summer 1980)[b] | | | | | | 1954 Kautz mudflow (%F)[c] | | 1977 Nooksack Cirque (%F)[d] | | |
| | Mud | | Blowdown | | Singe | | | | | | |
	%C	%F	%C	%F	%C	%F	Under snags	In open	Eroded till	Rubble meadows	Morainal mounds
Brassicaceae											
Cardamine oligosperma Nutt.	0	0	+	4	0	0	0	0	0	0	0
Saxifragaceae											
Boykinia elata (Nutt.) Greene	0	0	0.04	4	0	0	0	0	0	0	0
Heuchera glabra Willd.	0	0	0	0	0	0	0	0	0	50	0
Heuchera micrantha Dougl.	0	0	0.04	4	0	0	0	0	0	0	0
Mitella breweri Gray	0	0	+	4	0	0	1	0	0	0	0
Saxifraga arguta D. Don	0	0	0	0	0	0	0	0	0	13	0
Saxifraga ferruginea Grah.	0	0	0	0	0	0	0	0	43	50	0
Tiarella trifoliata L.	0	0	0	0	0	0	1	0	0	0	0
Tolmiea menziesii (Pursh) T. and G.	0	0	+	4	0	0	0	0	0	0	0

Grossulariaceae											
Ribes sp.	0	0	0	4	0	0	0	0	0	0	0
Rosaceae											
Aruncus sylvester Kostel.	0	0	+	4	0.50	13	0	0	0	25	0
Luetkea pectinata (Pursh) Kuntze	0	0	0	0	0	0	0	0	86	63	67
Rubus leucodermis Dougl.	0	0	0.21	4	0	0	1	0	0	0	0
Rubus parviflorus Nutt.	1.43	14	0.17	13	0.38	25	0	0	0	38	0
Rubus spectabilis Pursh	0.71	14	0.13	9	0	0	0	0	0	38	0
Rubus ursinus Cham. and Schlecht.	0	0	+	4	0	0	0	0	0	0	0
Sorbus sitchensis Roemer	0	0	0	0	0	0	0	0	14	38	0
Spiraea betulifolia Pall.	0	0	0	0	0	0	0	0	43	50	33
Fabaceae											
Lupinus latifolius Agardh[f]	0	0	+	4	1.25	38	0	0	29	13	7
Oxalidaceae											
Oxalis trilliifolia Hook.	0	0	0.09	4	0.38	13	0	0	0	0	0
Aceraceae											
Acer macrophyllum Pursh	0	0	+	4	0	0	0	0	0	0	0

APPENDIX 4.B
SPECIES LIST BY DISTURBANCE TYPE SHOWING MEAN PERCENTAGE OF COVER (%C) AND RELATIVE FREQUENCY (%F)[a](continued)

| Species | Mount St. Helens (Summer 1980)[b] | | | | | | 1954 Kautz mudflow (%F)[c] | | 1977 Nooksack Cirque (%F)[d] | | |
| | Mud | | Blowdown | | Singe | | | | | | |
	%C	%F	%C	%F	%C	%F	Under snags	In open	Eroded till	Rubble meadows	Morainal mounds
Polemoniaceae											
Phlox diffusa Benth.	0	0	0	0	0	0	0	0	0	25	0
Hydrophyllaceae											
Hydrophyllum fendleri (Gray) Heller	0	0	0.09	4	0	0	0	0	0	0	0
Phacelia hastata Dougl.	0	0	+	4	0	0	0	0	0	0	0
Lamiaceae											
Stachys cooleyae Heller[f]	0	0	0.09	4	0.25	13	0	0	0	0	0
Scrophulariaceae											
Mimulus guttatus DC.[f]	0	0	0.09	4	0	0	0	0	0	0	0
Mimulus lewisii Pursh	0	0	0	0	0.25	0	0	0	0	13	0
Penstemon cardwellii Howell	0	0	+	4	0	0	0	0	0	0	0
Penstemon davidsonii Greene	0	0	0	0	0	0	0	0	29	50	13

Species									
Penstemon procerus Dougl.	0	0	0	0	0	0	0	0	13
Rubiaceae									
Galium sp.	0	0	0	0	0	0	0	0	0
Caprifoliaceae									
Sambucus racemosa L.	0	0	0	0	0	1	0	14	50
Valerianaceae									
Valeriana sitchensis Bong.	0	0	0	0	0	+	0	0	38
Campanulaceae									
Campanula rotundifolia L.	0	0	0	0	0	0	0	14	100
Asteraceae									
Achillea millefolium L.	0	0	0	0	0	0	0	0	63
Adenocaulon bicolor Hook.	0	0	0	0.13	13	0	0	0	0
Agoseris aurantiaca (Hook.) Greene	0	+	4	0	0	56	0	0	0
Anaphalis margaritacea (L.) B. and H.	0	0.17	17	0.75	25	100	83	71	50
Arnica amplexicaulis Nutt.	0	0	0	0	0	0	0	0	13
Artemisia ludoviciana Nutt.	0	0	0	0	0	0	0	0	13

APPENDIX 4.B
Species List by Disturbance Type Showing Mean Percentage of Cover (%C) and Relative Frequency (%F)[a] (continued)

Species	Mount St. Helens (Summer 1980)[b]						1954 Kautz mudflow (%F)[c]		1977 Nooksack Cirque (%F)[d]		
	Mud		Blowdown		Singe		Under snags	In open	Eroded till	Rubble meadows	Morainal mounds
	%C	%F	%C	%F	%C	%F					
Aceraceae cont'd											
Acer circinatum Pursh	0.28	14	0.04	4	0	0	0	0	0	0	0
Violaceae											
Viola adunca Sm.	0	0	+	4	0	0	0	0	0	13	0
Viola orbiculata Geyer	0	0	+	4	0	0	0	0	0	0	0
Onagraceae											
Circaea alpina L.	0	0	+	4	0	0	0	0	0	0	0
Epilobium alpinum L.	0	0	0	0	0	0	0	0	14	50	0
Epilobium angustifolium L.[f]	0	0	5.70	50	0.63	13	100	100	0	50	0
Epilobium glaberrimum Barbey	0	0	0.09	4	0	0	78	33	0	0	0
Epilobium sp.	0	0	0	0	0	0	0	0	14	50	0

Taxon											
Apiaceae											
Heracleum lanatum Michx.	0	0	+	4	0	0	0	0	0	0	0
Lomatium sp.	0	0	0	0	0	0	0	0	0	13	0
Osmorhiza occidentalis (Nutt.) Torr.	0	0	0	0	0	0	0	0	0	25	0
Cornaceae											
Cornus canadensis L.	0	0	+	4	0.38	13	3	0	0	0	0
Ericaceae											
Cassiope mertensiana (Bong.) G. Don	0	0	0	0	0	0	0	0	14	13	0
Chimaphila menziesii (R. Br.) Spreng.	0	0	+	4	0	0	0	0	0	0	0
Cladothamnus pyroliflorus Bong.	0	0	0	0	0	0	0	0	29	0	73
Menziesia ferruginea Smith	0	0	+	4	0.25	25	0	0	0	0	7
Phyllodoce empetriformis (SW.) D. Don	0	0	0	0	0	0	0	0	86	13	40
Vaccinium membranaceum Dougl.	0	0	0	0	0	0	0	0	29	0	7
Vaccinium ovalifolium Smith	0	0	+	4	0.25	25	0	0	0	0	7
Vaccinium parvifolium Smith	0	0	0	0	0	0	11	0	0	0	0

SPECIES LIST BY DISTURBANCE TYPE SHOWING MEAN PERCENTAGE OF COVER (%C)
AND RELATIVE FREQUENCY (%F)[a] (continued)

Species	Mount St. Helens (Summer 1980)[b]						1954 Kautz mudflow (%F)[c]		1977 Nooksack Cirque (%F)[d]		
	Mud		Blowdown		Singe		Under snags	In open	Eroded till	Rubble meadows	Morainal mounds
	%C	%F	%C	%F	%C	%F					
Asteraceae cont'd											
Chrysanthemum leucanthemum L.[f]	0	0	0	0	0	0	13	25	0	0	0
Cirsium arvense (L.) Scop.[f]	0	0	1.09	13	1.63	38	44	0	0	0	0
Cirsium edule Nutt.	0	0	0	0	0	0	0	0	0	63	0
Crepis sp.	0	0	0	0	0	0	1	0	0	0	0
Erigeron peregrinus (Pursh) Greene	0	0	0	0	0	0	0	0	29	50	0
Hieracium albiflorum Hook.	0	0	+	4	0.13	13	100	50	0	13	7
Hieracium gracile Hook.	0	0	0	0	0	0	0	0	29	50	0
Hypochaeris radicata L.[f]	0	0	+	4	0.38	25	100	83	0	0	0
Luina hypoleuca Benth.	0	0	0	0	0	0	0	0	14	13	0
Petasites frigidus (L.) Fries	0	0	0	0	0	0	6	0	0	0	0
Senecio jacobaea L.[f]	0	0	+	4	0.63	13	0	0	0	0	0
Senecio triangularis Hook.	0	0	+	4	0.38	25	56	33	0	0	0

Species											
Sonchus sp.	0	0	0	0	1	0	0	0	0	0	0.14
Taraxacum officinale Weber	0	0	0	0	1	0	0	4	+	0	0
Juncaceae											
Juncus drummondii E. Meyer	20	63	57	0	0	0	0	0	0	0	0
Juncus mertensianus Bong.	0	13	43	0	0	0	0	0	0	0	0
Juncus sp.	0	0	0	0	1	0	0	0	0	0	0
Luzula divaricata Wats.	0	0	0	0	0	0	0	4	+	0	0
Luzula hitchcockii Hamet-Ahti[f]	7	13	71	0	0	0	0	0	0	0	0
Luzula spicata (L.)	0	0	0	0	0	13	+	0	0	0	0
Luzula sp. DC.	0	0	0	0	2	0	0	0	0	0	0
Cyperaceae											
Carex mertensii Prescott[f]	0	25	0	0	0	38	1.50	4	+	14	0.14
Carex pachystachya Cham.[f]	0	38	86	0	0	13	0.25	4	+	0	0
Carex spectabilis Dewey	0	100	71	0	0	0	0	0	0	0	0
Carex spp.	0	0	0	0	12	0	0	0	0	0	0

SPECIES LIST BY DISTURBANCE TYPE SHOWING MEAN PERCENTAGE OF COVER (%C) AND RELATIVE FREQUENCY (%F)[a] (continued)

Species	Mount St. Helens (Summer 1980)[b]						1954 Kautz mudflow (%F)[c]		1977 Nooksack Cirque (%F)[d]		
	Mud		Blowdown		Singe		Under snags	In open	Eroded till	Rubble meadows	Morainal mounds
	%C	%F	%C	%F	%C	%F					
Poaceae											
Aira praecox L.	0	0	0	0	0	0	1	0	0	0	0
Agrostis diegoensis Vasey[f]	0	0	+	4	0	0	0	0	0	0	0
Agrostis variabilis Rydb.	0	0	0	0	0	0	0	0	0	13	0
Agrostis spp.	0	0	0	0	0	0	0	0	57	38	0
Calamagrostis canadensis (Michx.) Beauv.	0	0	0	0	0	0	0	0	14	75	7
Calamagrostis sesquiflora (Trin.) Kawano	0	0	0.09	4	0	0	0	0	0	0	0
Deschampsia atropurpurea (Wahl.) Scheele	0	0	0	0	0	0	0	0	86	63	13
Deschampsia elongata (Hook.) Munro[f]	0	0	0	0	+	4	0	0	0	13	0
Elymus glaucus Buckl.[f]	0	0	0	0	0.13	13	0	0	14	100	0
Elymus sp.	0	0	0	0	0	0	1	0	0	0	0

Species										
Holcus lanatus L.	0	0	0	0	0	1	0	0	0	0
Phleum alpinum L.	0	0	0	0	0	0	0	14	25	0
Poa sp.[f]	0	0	13	0.25	13	0	0	0	50	0
Trisetum spicatum (L.) Richter[f]	0	0.13	13	+	0	1	0	14	38	0
Typhaceae										
Typha latifolia L.	0	+	4	0	0	0	0	0	0	0
Liliaceae										
Clintonia uniflora (Schult.) Kunth.	0	0.09	4	0	0	1	0	0	0	0
Lilium columbianum Hanson	0	0	0	0	0	1	0	0	0	0
Smilacina stellata (L.) Desf.	0	0	0	+	13	0	0	0	0	0
Veratrum viride Ait.	0	0.22	13	0	0	0	0	0	13	0

[a]Taxonomic authority is Hitchcock et al. (1973).
[b]Includes only plants found in plots.
[c]From Frehner 1957.
[d]From Oliver et al. 1979.
[e]The "+" symbol refers to plants that were present in trace amounts.
[f]Plants that flowered and set seed within the blast zone in the summer of 1980.

APPENDIX 4.C
VASCULAR PLANT SPECIES INCLUDING DISTURBANCE TYPE CREATED BY THE MAY 18, 1980, LATERAL BLAST OF MOUNT ST. HELENS[a]

Abies amabilis (Dougl.) Forbes, b, d, m, s
Abies procera Rehder, b, d, n[c]
Acer circinatum Pursh, b, d, m, s, n
Acer glabrum Torr. var. *douglasii* (Hook.) Dippel, b
Acer macrophyllum Pursh, b, d, m, s
Achillea millefolium L., m
Achlys triphylla (Smith) DC., b, s
Actaea rubra (Ait.) Willd., b, s
Adenocaulon bicolor Hook., b, s
Adiantum pedatum L., s
Agrostis diegoensis Vasey, b, d, m, s
Agrostis scabra Willd., b, d, m, s
Agrostis tenuis Sibth., m
Agrostis thurberiana Hitchc., d, m
Agrostis spp., m
Aira praecox L., m
Alnus rubra Bong., b, d, m, s
Alnus sinuata (Regel) Rydb., b, n
Alopecurus geniculatus L., d, m
Anaphalis margaritacea (L.) B. & H., b, d, m, s, n
Anthoxanthum odoratum L., b
Arctostaphylos uva-ursi (L.) Spreng., b
Arenaria serpyllifolia L., b, d, s
Arnica diversifolia Greene, b
Aruncus sylvester Kostel., b
Aster ledophyllus Gray, b
Athyrium filix-femina (L.) Roth., b, d, m, s, n
Berberis aquifolium Pursh, b, s
Berberis nervosa Pursh, b, d, m, s
Bidens cernua L., m
Blechnum spicant (L.) Roth., b, d, m, s
Botrychium lanceolatum (Gmel.) Angstr., b
Botrychium multifidum (Gmel.) Trevis., m
Boykinia elata (Nutt.) Greene, b
Bromus spp., b, d
Bromus carinatus H. & A. var. *carinatus*, b
Calamagrostis sesquiflora (Trin.) Kawano, b
Caltha biflora D.C., b
Campanula scouleri Hook., b
Cardamine angulata Hook., b, s
Cardamine oligosperma Nutt., d
Carex disperma Dewey, b, s
Carex mertensii Prescott, b, d, m, s
Carex pachystachya Cham., b, d, s, n
Carex spectabilis Dewey, d, n
Carex spp., b, d, m

APPENDIX 4.C
VASCULAR PLANT SPECIES INCLUDING DISTURBANCE TYPE CREATED BY
THE MAY 18, 1980, LATERAL BLAST OF MOUNT ST. HELENS[a] (continued)

Cerastium arvense L., b, d, s
Chimaphila menziesii (R. Br.) Spreng., b, d
Chrysanthemum leucanthemum L., b, s
Circaea alpina L., b, d
Cirsium arvense (L.) Scop., b, d, m, s
Cirsium vulgare (Savi) Tenore, b, d, m, s
Clintonia uniflora (Schult.) Kunth, b
Collomia debilis (S. Wats.) Greene var. *larsenii* (Gray) Brand, n
Collomia heterophylla Hook., b, d, m, s
Cornus canadensis L., b, d, s, n
Corydalis scouleri Hook., b, s
Cryptogramma crispa (L.) R. Br., n
Cystopteris fragilis (L.) Bernh., d, m
Dactylis glomerata L., b, d, m, s
Danthonia intermedia Vasey, d
Daucus carota L., b
Deschampsia danthonioides (Trin.) Munro, b, d, s
Deschampsia elongata (Hook.) Munro, b, d
Dianthus armeria L., b
Dicentra formosa (Andr.) Walp., b, d, s
Digitalis purpurea L., b, d, m, s
Disporum smithii (Hook.) Piper, b, s
Draba verna L., d
Elymus glaucus Buckl., b, d, m, s
Epilobium alpinum L., n
Epilobium angustifolium L., b, d, m, s
Epilobium glaberrimum Barbey, b, d, m, s
Epilobium latifolium L., n
Epilobium minutum Lindl., d
Epilobium paniculatum Nutt., b, m, s
Equisetum fluviatile L., b, d, m, s
Equisetum palustre L., b, d, m, s
Erigeron annuus (L.) Pers., m
Eriogonum pyrolifolium Hook., var. *coryphaeum* T. & G., b, n
Erythronium grandiflorum Pursh, b, s
Erythronium montanum Wats, b, s
Festuca arundinacea Schreb., a[b]
Festuca idahoensis Elmer, b
Festuca rubra L., b, d, m, s[b]
Galium asperrimum Gray, d
Galium triflorum Michx., b, d, m, s
Gaultheria shallon Pursh, b, d, m, s
Geum macrophyllum Willd., m
Glyceria elata (Nash) Jones, b, m
Gnaphalium microcephalum Nutt., b, s
Gratiola neglecta Torr., b

APPENDIX 4.C

VASCULAR PLANT SPECIES INCLUDING DISTURBANCE TYPE CREATED BY
THE MAY 18, 1980, LATERAL BLAST OF MOUNT ST. HELENS[a] (continued)

Grindelia integrifolia DC., b, s
Gymnocarpium dryopteris (L.) Newm., b, s
Heracleum lanatum Michx., b
Heuchera micrantha Dougl., b
Hieracium albiflorum Hook., b, d, s
Holcus lanatus L., b, d, m, s
Holodiscus discolor (Pursh) Maxim., b
Hydrophyllum fendleri (Gray) Heller, b, d, s
Hypochaeris glabra L., b, m, s
Hypochaeris radicata L., b, d, m, s
Hypericum perforatum L., m
Juncus bufonius L., b, d, m
Juncus effusus L., b, d, m, s
Juncus ensifolius Wikst., b, d, m, s
Juncus regelii Buch., b, d, m, s
Lactuca muralis (L.) Fresen, b, m, s
Lactuca serriola L., b, d, m, s
Ligusticum grayi Coult. & Rose, b
Lilium columbianum Hanson, b
Linnaea borealis L., m
Lolium multiflorum Lam., b, d, m, s[c]
Lolium perenne L., b, d, m, s[c]
Lolium temulentum L., b, d, m, s[b]
Lomatium martindalei Coult. & Rose, n
Lotus corniculatus L., b, d[c]
Lotus micranthus Benth., b
Lotus pedunculatus Cav.[b]
Lotus purshiana (Benth.) Clements & Clements, b, d
Luetkea pectinata (Pursh) Kuntze., n
Lupinus albicaulis Dougl., d[b]
Lupinus lepidus Dougl., var. *lobbii* (Gray) Hitchc., b, d, n
Lupinus latifolius Agardh, b, d, n
Lupinus rivularis Dougl., d
Luzula divaricata Wats., b, d, m, s, n
Luzula campestris (L.) DC., b, s, m, n
Luzula parviflora (Ehrh.) Desv., m
Lysichitum americanum Hulten & St. John, b
Madia sativa Mol., b, s
Maianthemum dilatatum (Wood) Nels. & Macbr., b, s
Matricaria matricarioides (Less.) Porter, b
Menziesia ferruginea Smith., b, s
Mimulus guttatus DC., b, d
Mitella breweri Gray, b
Montia parvifolia (Moc.) Greene var. *flagellaris* (Bong.) Hitchc., b, d, m, s, n
Montia perfoliata (Donn) Howell, b
Montia sibirica (L.) Howell, b, d, m, s

APPENDIX 4.C

VASCULAR PLANT SPECIES INCLUDING DISTURBANCE TYPE CREATED BY THE MAY 18, 1980, LATERAL BLAST OF MOUNT ST. HELENS[a] (continued)

Myosotis arvensis (L.) Hill, b, m, s
Navarretia squarrosa (Esch.) H. & A., b
Oenanthe sarmentosa Presl., b, d
Oplopanax horridum (Smith) Miq., b, s
Osmorhiza chilensis H. & A., b
Oxalis trilliifolia Hook., b, s
Penstemon cardwellii Howell, b
Penstemon procerus Dougl., b
Petasites frigidus (L.) Fries, b, d, m, s, n
Phacelia hastata Dougl., var *leptosepala* (Rydb.) Cronq., b
Phacelia nemoralis Greene, b
Phleum alpinum L., b, n
Phleum pratense L., b, d, m, s[d]
Physocarpus capitatus (Pursh) Kuntze, m
Picea sitchensis (Bong.) Carr., m
Pinus contorta Dougl., d[b]
Pinus monticola Dougl., d[b]
Plantago lanceolata L., b, s
Plantago major L., b, s
Poa annua L., b
Poa nervosa (Hook.) Vasey, b, d, m, s
Poa pratensis L., b, d, s
Poa spp., b, d
Polygonum minimum Wats., b, n
Polygonum newberryi Small, b, n
Polygonum persicaria L., s, b, d
Polystichum lonchitis (L.) Roth, n
Polystichum munitum (Kaulf.) Presl var. *munitum*, b, d, m, s
Populus trichocarpa T. & G., b, d, m, s
Potentilla biennis Greene, d
Prunella vulgaris L., b
Prunus emarginata (Dougl.) Walp., b, d, m, s
Pseudotsuga menziesii (Mirbel) Franco. var. *menziesii*, b, d, m, s, n[b]
Pteridium aquilinum (L.) Kuhn., b, d, m, s
Pyrola secunda L., d
Ranunculus repens L., b
Rhamnus purshiana DC., b
Ribes bracteosum Dougl., m
Ribes lacustre (Pers.) Poir., b
Ribes sanguineum Pursh, b
Ribes watsonianum Koehne, b
Rosa gymnocarpa Nutt., b
Rosa nutkana Presl, b, m, s
Rosa pisocarpa Gray, b, s
Rubus discolor Weihe & Nees, b, s
Rubus idaeus L., b, d

APPENDIX 4.C
VASCULAR PLANT SPECIES INCLUDING DISTURBANCE TYPE CREATED BY THE MAY 18, 1980, LATERAL BLAST OF MOUNT ST. HELENS[a] (continued)

Rubus laciniatus Willd., b, s
Rubus lasiococcus Gray, b
Rubus leucodermis Dougl., b, d, m, s
Rubus nivalis Dougl., b
Rubus parviflorus Nutt., b, d, m, s
Rubus pedatus J. E. Smith, b
Rubus spectabilis Pursh, b, d, m, s, n
Rubus ursinus Cham. & Schlecht., b, d, m, s
Rumex acetosella L., b, d, m, s
Rumex crispus L., m
Rumex obtusifolius L., b, d, m
Salix lasiandra Benth., b, d, m
Salix scouleriana Barratt, b, d, m, s, n
Salix sitchensis Sanson, b, d, m, s[c]
Salix sp. L., b, d, m, s
Sambucus racemosa L., b, d, m, s, n
Saxifraga ferruginea Grah., b, d, n
Scirpus microcarpus Presl, b, d, s
Senecio jacobaea L., b, d, m, s, n
Senecio sylvaticus L., b, d, m, s, n
Senecio triangularis Hook., b, m, s
Senecio vulgaris L., m
Sitanion hystrix (Nutt.) Smith, b, n
Smilacina stellata (L.) Desf., b, m, s
Sonchus arvenis L., b, d, s
Solidago canadensis L., b
Sorbus sitchensis Roemer var. *grayi* (Wenzig) Hitchc., b
Spraguea umbellata Torr. var. *caudicifera* Gray, b, n
Spergula arvensis L., b
Spiraea douglasii Hook., m
Stachys cooleyae Heller, b, s
Stellaria calycantha (Ledeb.) Bong., b, m, s
Stellaria longipes Goldie, m
Streptopus amplexifolius (L.) DC., b
Taraxacum offinale Weber, b
Thuja plicata Donn., m
Tiarella trifoliata L., b, s
Tolmiea menziesii (Pursh) T. & G., b, d, s
Trautvetteria caroliniensis (Walt.) Vail., b
Trientalis latifolia Hook., b, s, d
Trifolium microcephalum Pursh, d
Trifolium pratense L., b, d, m, s[c]
Trifolium repens L., b, d, m, s[c]
Trillium ovatum Pursh, b, s, n
Trisetum cernuum Trin., b, s

APPENDIX 4.C

VASCULAR PLANT SPECIES INCLUDING DISTURBANCE TYPE CREATED BY THE MAY 18, 1980, LATERAL BLAST OF MOUNT ST. HELENS[a] (continued)

Trisetum spicatum (L.) Richter, b, n
Tsuga heterophylla (Raf.) Sarg., d
Tsuga mertensiana (Bong.) Carr., b
Typha latifolia L., b, d, m, s
Urtica dioica L. var. *lyallii* (Wats.) Hitchc., b
Vaccinium deliciosum Piper, b, n
Vaccinium membranaceum Dougl., b, d, n
Vaccinium ovalifolium Smith, b, s
Vaccinium parvifolium Smith, b, s
Valeriana sitchensis Bong., b, n
Vancouveria hexandra (Hook.) Morr. & Dec., b, d, s, n
Veratrum viride Ait., b
Veronica anagallis-aquatica L., b, d, m, s
Veronica serpyllifolia L., b
Vicia villosa Roth[b]
Viola glabella Nutt., b, s
Viola orbiculata Geyer, b, s
Xerophyllum tenax (Pursh) Nutt., b, s

[a]Most of the species (93%) were found from the summer of 1980 through the summer of 1982; the remainder were found in 1984. Letters following the Latin binomials refer to disturbance type in which the species was located: b = blowdown, d = debris slide, m = mudflow, s = scorch forests, and n = north and northwest slope of Mount St. Helens.

[b]Planted since eruption, not residual.

[c]Planted since eruption and also residual.

5

Initial Vegetation Recovery on Subalpine Slopes of Mount St. Helens, Washington

Roger del Moral and L. C. Bliss

ABSTRACT

The recovery of subalpine plants surviving the May 18, 1980, eruption of Mount St. Helens was observed during 8 days from July to September, 1980. Because of the protection afforded by the snow pack at higher elevations, much herbaceous vegetation on the western and southwestern slopes of the volcanic cone was initially unaffected by the devastating eruption. Where 6–9 cm of tephra was subsequently deposited, 70–90% of the individuals were able to penetrate the layer during the 1980 growing season. Grasses, however, appeared more susceptible to burial and suffered greater mortality than did most other herbs. In areas where thin layers of mud buried vegetation but were subsequently eroded, regrowth of vegetation occurred rapidly. Plant recovery was observed up to ~2300 m on the slopes of the volcano. Where all vegetation was killed, recovery will require both soil development and colonization from less devastated areas. Results of seed germination trials suggest that many species are capable of germinating in harsh substrates.

INTRODUCTION

Little is known about the subalpine vegetation of Mount St. Helens prior to May 18, 1980, but the existing evidence suggests that it was immature, having developed on substrates of very recent origins (F. Ugolini, personal communication 1980, Crandell and Mullineaux 1978). Typically, vegetation was open and sparse (A. R. Kruckeberg, personal communication 1980) and of limited species richness (St. John 1976; Lawrence 1941). In 1938, the treeline on the north slope was at least 800 m lower than that of neighboring volcanic cones (Lawrence 1941). Apparently, frequent eruptions of Mount St. Helens

over the past 4500 yr have repeatedly retarded vegetation development. Crandell and Mullineaux (1978) report 5 major events of
pumice deposition, at least 14 episodes of mudflows on various flanks,
14 episodes of pyroclastic flows, and 8 major lava flows. Isolation from
other alpine floras have combined with chronic disturbance and a
porous substrate to limit the species richness of the subalpine flora of
this volcano.

In this chapter, we describe the damage sustained by subalpine
herbaceous vegetation following the eruptions of late spring and
summer of 1980, its recovery by September 1980 at selected sites above
1250 m at and above treeline, and offer a prognosis for colonization of
the devastated areas. We examined sites on all but the north flank of
the cone.

MATERIALS AND METHODS

FIELD SAMPLING

During four reconnaissance trips during July and August 1980, the
general extent of damage on the volcano was assessed. During an
intensive 4-day research effort in early September, subalpine herbaceous vegetation within 42 permanently marked 250-m² circular plots
was sampled. Within each plot, plant cover was estimated visually in six
20 × 50-cm quadrats located on each of four marked radii.

No plots were established in the upper pyroclastic zone above
1200 m where aerial and a brief ground inspection revealed no vegetation in 1980. One plot was established on Abraham Plains at 1340 m
elevation. The following transects were established: Butte Camp at
1350 and 1550 m elevation on tephra deposits overlaying a 300–
500-yr-old mudflow (F. Ugolini, personal communication 1980), on
steep slopes above 1980 mudflows, and on these mudflows; near Pine
Creek at 1280 m on the margin of the Pine Creek mudflow generated
by the melting of the Shoestring Glacier; at 1370 m on a thin mudflow;
and at 1530 m on a thicker part of the same mudflow. Scattered plots
also were placed on a dry ridge covered by tephra at 1570 m above the
South Fork of the Toutle River and on a mesic ridge at 1580 m at the
edge of the blast zone.

VEGETATION PATTERN ANALYSIS

The 38 plots with vegetation were analyzed by detrended correspondence analysis (Hill and Gauch 1980). This indirect ordination tech-

nique seeks to arrange samples according to their floristic relationships in such a way that vegetation response to environmental gradients can be revealed. The method is relatively unaffected by the stresses created by large floristic differences among the samples (Gauch 1982), although differences in absolute cover can make interpretations difficult (del Moral 1983). Statistical relationships between axes are removed, and the axes are scored as a measure of floristic distance (β diversity) that is constant throughout. Mean absolute cover percentages from the 24 quadrats in each plot were converted to octave scores to reduce the contribution of abundant species, and the importance of rare species (fewer than five occurrences) was downweighted (Hill 1979) to facilitate interpretations. Floristic differences resulting from geography and large differences in total plot cover percentages can stress this method, but the emergent patterns were at least as clear as those obtained from reciprocal averaging, principal components analysis, and polar ordination.

As an aid to characterizing the vegetation structure in these samples, we calculated the Shannon–Weiner index (H')

$$H' = - \sum_{i=1}^{n} P_{ij} \ln P_{ij}$$

where P_{ij} is the cover proportion of the ith species in the jth sample and n is the number of species in the sample. This index increases as the proportions become more nearly equal and as the number of species increases.

PRELIMINARY GERMINATION TRIALS

The following raw volcanic materials were collected as substrates for germination tests of selected species: pyroclastic powder obtained from 1500 m elevation on the north slope of the crater, mud from Pine Creek, mud from Butte Camp, and coarse tephra from Butte Camp. Twenty stratified seeds were placed in each of two petri plates for each substrate and were irrigated with distilled water. Plates were sealed and placed in a high illumination growth chamber on a 16-hr photoperiod with a 20°C day and 10°C night temperature cycle. Germination was recorded daily, and plants were irrigated as needed.

RESULTS

GENERAL VEGETATION PATTERNS

We recognized five categories of volcanic damage in the subalpine zone resulting from eruptions in 1980:

1. A large area north of the crater was destroyed by the combined effects of landslides, a debris flow, and pyroclastic flows (Rosenfeld 1980).

2. The eastern flank, including Abraham Plains, received hot pumice and mudflows that destroyed all vegetation. These plains are covered by pumice and mud of varied composition on which no vegetation was found in September, 1980.

3. On the southern and western slopes, small mudflows resulting from deglaciation scoured many high valleys and deposited fine mud of varying thicknesses on upper ridges and lower slopes.

4. Tephra ejected by the eruptions of May 18, May 25, and June 12, and parts of some mudflows were deposited over snow that provided protection to the dormant vegetation. These deposits on the southwestern, southern, and eastern flanks of the volcano subsequently inhibited the emergence of most species.

5. Sites on the western edge of the blast zone were protected from direct blast effects by snow and were sufficiently close to the crater to receive little tephra. Although scattered *Pinus contorta* (nomenclature follows that of Hitchcock and Cronquist 1973) individuals were killed by the heat blast, herbaceous vegetation appeared unscathed and normal in September, 1980.

The permanent plots established in September, 1980, reflect the range of volcanic effects on the vegetation observed up to that time. Figure 5.1 displays those plots containing any vegetation as determined by detrended correspondence analysis (DCA). Axis 1 is a moisture gradient, from relatively dry sites on the left to mesic or protected sites on the right. Axis 2 reflects the degree of mud or tephra deposition. Neither axis is "pure" since the vegetation response reflected by the analysis combines long-term environmental effects (e.g., soil moisture, snow deposition, and substrate differences) with the recent volcanic effects.

The thick mudflow plots at Pine Creek had little cover but moderate species richness. They are confined to the upper left of Fig. 5.1. *Eriogonum pyrolifolium* and *Lupinus lepidus* var. *lobbii* were widely scattered at these locations. Plots on similarly exposed substrates at

Butte Camp are situated to the extreme left of Fig. 5.1. These plots
sampled mudflow margins and had more cover and greater species
richness, including *Polygonum newberryi* and *Juncus parryi* as characteristic species. Plots on thin mudflows and on steep slopes with few
species are in the lower left of Fig. 5.1. *Agrostis diegoensis* is the most
common species in these plots. The central portion of Fig. 5.1 is
occupied by those Butte Camp plots that were primarily affected by
tephra falling on the snow pack. Hence, little damage occurred to
vegetation, although the total cover was greatly reduced. The lower
midrange of Fig. 5.1 contains communities from sites exposed to direct
tephra deposition. *Aster ledophyllus* is a dominant species in these sites,
where woody species suffered substantial damage. Two sets of plots
are represented in the far right portion of Fig. 5.1. The upper set
consists of Pine Creek plots covered only by a thin layer of mud, much
of which had been eroded away by September, 1980. *Lupinus latifolius*

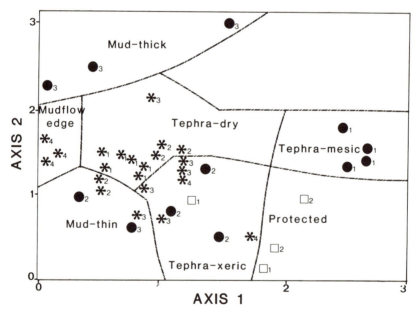

FIG. 5.1   Detrended correspondence analysis of plots in the subalpine zone
of Mount St. Helens. Axis 1 and 2 are the first two axes identified by this
method. * = Butte Camp: 1. low elevation tephra; 2. high elevation tephra;
3. steep slopes; 4. mudflow edges. ● = Pine Creek: 1. protected with tephra
and some mud; 2. thin mudflow; 3. thick mudflow. □ = South Fork Toutle
ridges: 1. dry ridge; 2. mesic ridge.

and *Xerophyllum tenax* characterize this vegetation. The lower set derives from the nearly unaffected plots on the ridges above the South Fork Toutle that are rich in species and usually high in percentage of cover.

BUTTE CAMP

The structural features of the vegetation are shown in Table 5.1 (Butte Camp), Table 5.2 (Pine Creek), and Table 5.3 (ridges on the western flank). In each table, the number of plots in each category, mean percentage of plant cover, mean species richness, mean Shannon–Weiner (information theory) index (H'), and a truncated species list showing frequency and mean percentage of cover (cover in each plot was estimated from 24 0.1-m^2 samples) is presented.

The effects of tephra falling on dormant, snow-covered, herbaceous vegetation can be investigated using the Butte Camp transects

FIG. 5.2   Upper transect at Butte Camp, showing scattered vegetation in the foreground and nearly total destruction due to mudflows in the distance (September 1980).

(Table 5.1; Fig. 5.2). The dominants include *Agrostis diegoensis, Eriogonum pyrolifolium, Lupinus lepidus, Lomatium martindalei*, and *Phlox diffusa*. Excavation of tephra down to the original soil surface (a mudflow several centuries old) revealed that some plants failed to penetrate deposits of as little as 10 cm; however, dead plant shoots accounted for only 27 and 20% of the total plant cover in samples at the 1350 and 1550 m levels, respectively. Most mortality occurred among graminoid species. Live plant coverage in early September in the lower and upper transects was 16.1 (±10.0) and 18.7 (±9.8)%, respectively, indicating a general sparseness of plants on these relatively undisturbed upper slopes. These plants had emerged in June through 6−9 cm of tephra. Although many of them flowered, seed set was abnormally low compared to that of other subalpine areas in Washington in 1980.

Aboveground net annual production, determined by harvests ($n = 7$), averaged 29.8 (±14.9) g/m^2 at the lower site and 86.2 (±58.5) g/m^2 at the upper site in randomly sampled 1—m quadrats. Plant cover in these plots averaged 13.5 (±5.0) and 37.7 (±17.4)%, respectively. The latter estimates for cover and the associated dry matter production are considerably higher than the average cover and implied production of the high Butte Camp plots.

Where the snow pack was thinner, woody species including *Phyllodoce empetriformis, Phlox diffusa*, and *Arctostaphylos nevadensis* were substantially damaged by hot tephra. Scorched leaves and stems appeared dead in July and failed to recover by September.

The steeper slopes had less vegetative cover than other sites, but they were also buried under less tephra (usually <5 cm) because much of it was removed with the melting snow. Due to the greater heterogeneity of these sites, greater species richness was encountered. Such slopes may function as sources of disseminules for the colonization of nearby mudflows. There was virtually no vegetation on these mudflows, and even 10 cm appeared a sufficient mudflow depth to suppress resprouts of buried plants. Only *Lupinus lepidus* and *Eriogonum pyrolifolium* were found on mudflows, and these were presumably derived from surviving roots washed down with the flows. Both species are characteristic of chronically unstable substrates and can penetrate relatively unconsolidated material of substantial thickness. Plots on mudflow margins were included to identify possible species that could successfully colonize these flows. *Polygonum newberryi, Aster ledophyllus*, and *Agrostis diegoensis* appear most able to penetrate the thin marginal deposits. On the basis of their proximity and relatively high seed set in 1980, it is predicted that these five species will be the main early invaders of stabilized mudflows.

PINE CREEK: A HIGH ELEVATION MUDFLOW

At Pine Creek the melting of the Shoestring Glacier created a mudflow that swept the creek bed for several kilometers. At a study site between 1280 and 1530 m on a ridge separating Pine Creek from Muddy River, the mud washed over the creek bank and extended as a relatively thin layer burying, rather than removing, the vegetation. This layer was only a few centimeters thick at tree line (1250 m), but reached over 25 cm in thickness at 1500 m elevation.

Plant cover and species richness were substantially affected by this mud deposition (Table 5.2). At the leading edge of the flow, the thin unconsolidated mud and tephra were easily broached by many buried species. Plant cover, although variable, was comparable to that of Butte Camp, but species richness was reduced. The species present suggested a more mesic preeruption community and the possibility that several species had failed to emerge at this site. The thin layer of mud had a major impact on vegetation, and plant cover was substantially reduced. Plants (including *Agrostis diegoensis*) were confined primarily to small erosion channels in the mudflow exposing the underlining pumice layer by the end of the summer (Fig. 5.3). *Aster ledophyllus* was the only species to consistently sprout through undisturbed mud, although *Lupinus lepidus* and *Polygonum newberryi* did so in several places. Other species required the erosion of mud for emergence. Of the ten species that occurred in four plots, eight were confined to areas where rill erosion had removed the mud. Fine-grained clay in the mud may have sufficiently sealed the surface, thereby limiting vegetation gas exchange and inhibiting recovery.

Still higher on the mudflow, plant cover was virtually nil and mean richness was low. *Eriogonum pyrolifolium* was the only species to occur in more than one plot, and all species were confined to deep erosion channels. Because of their abilities to survive mud deposition and to germinate on infertile soils, *Lupinus lepidus* and *Aster ledophyllus* may be major recolonizing species here in the future. The rate of vegetation recovery on this mudflow will depend on the rate of mud removal by erosion. Buried plants have been reported to survive for several years after the eruption at Katmai (Griggs 1919), and erosion would release surviving plants from suppression. During the winter of 1980−81, most mud was removed from this ridge leaving a surface of coarse pumice. Where mud had been moderately thick, however, vegetation was sparse and confined largely to the rills created in the summer of 1980. Short-term burial by fine mud may have caused substantial mortality at this site.

Community structure can be altered differentially by envi-

FIG. 5.3   At Pine Creek, vegetation was confined mainly to erosion channels. The primary exception was *Aster ledophyllus*, capable of growing through at
least 10 cm of mud. Plants shown here, primarily *Lupinus latifolius*, were
scorched by the August 7, 1980, eruption. (Photo taken August 20, 1980.)

ronmental impacts. Figure 5.4 shows the relationship between the
Shannon−Weiner index (H'), the percentage of cover, and the amount
of mud deposition at Pine Creek. At the foot of the deposit, the cover
was relatively high ($\bar{x} = 20.5\%$) and H' averaged 1.07 (± 0.17). The thin
mudflow substantially reduced the cover, but richness was high due to
the effects of rill erosion. Thus the sparse vegetation was relatively
evenly divided among the species and H' was high but variable at 1.19
(± 0.47). Thick mud reduced values of both the cover and H', but
because no semblance of a dominance hierarchy could be determined
at the sampling intensity employed, H' remained high except where
no species occurred. Under these conditions, the role of interspecific
competition in structuring vegetation is drastically reduced when fortuitous factors are present.

     Pine Creek was also exposed to a heat blast from the August 7,
1980, eruption. Plants directly exposed were severely scorched and

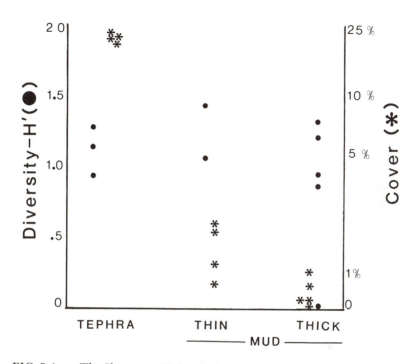

FIG. 5.4   The Shannon−Weiner index (H'), a measure of diversity, and the percentage of cover (log scale) on the Pine Creek mudflow. Tephra sites received a coarse aerial deposit and a very thin silt layer. The thin mudflow received 4−8 cm of silt, and the thick mudflow received >15 cm of silt.

usually dead to ground level. Had the May 18 eruption occurred in August, vegetation destruction on the volcano would have been far greater than it was.

RIDGE VEGETATION

On the western slope of the mountain, sites were established on two ridges. The first was outside the direct blast zone on a lava ridge that received heavy tephra deposits. Plant cover and species richness (Table 5.3) were similar to those of lower Butte Camp, but *Penstemon confertus*, *Agrostis diegoensis*, and *Achillea millefolium* var. *lanulosa* were the dominant species. These species survived because the snow pack protected them from searing heat. They will provide a source of seeds to colonize the denuded glacial valleys located below them. The rapidity with which some species adapt to tephra deposits is evidenced by *Collomia debilis*, which grew through 11 cm of tephra and had initiated adventitious roots from the buried stem tissue.

The second ridge site, located at 1580 m above the Talus Glacier valley, was just within the blast zone of the lateral eruption of May 18. The few *Pinus contorta* trees found here were killed by the scorching heat, but snow protected the dormant herbs. Little tephra fell at sites so close to the crater, and the surviving vegetation displayed vigorous growth (Fig. 5.5). Cover and species richness were maximum at this site, and dense vegetation extended up the western face to ~1800 m. Above this elevation, the slope was too steep and was unstable to support such vegetation (Fig. 5.5). This vegetation will also act as a primary source of disseminules for the denuded valleys and much of the upper debris flow that fills the valley of the North Fork of the Toutle River.

PHYSIOLOGICAL ADAPTATIONS

Considering the coarse and porous texture of the soils derived from pumice and tephra, their light brown surfaces, their low species richness, and their plant cover and plant productivity, it would be logical to assume that these volcanic slopes would have species with special morphological and physiological adaptations. Although graminoid species have extensive root systems for absorption, most of them suffered disproportionately from tephra deposition in 1980. This may be due to their limited ability to store food reserves in below-ground organs.

Aster ledophyllus, Lupinus lepidus, Eriogonum pyrolifolium, and *Polygonum newberryi* produce deep taproots with considerable storage tissue and exhibit rapid and vigorous shoot development. There was little evidence during the first summer that shoots of these species penetrating the tephra extended adventitious roots laterally into the new substrate. These plants have large dark green leaves and a diffuse growth form. Few cushion plants or other low growth form species are present. The morphological characteristics suggest a flora that is not unduly drought stressed. In fact, tephra-derived soils may hold considerable amounts of water below -1.0 mPa (Maeda et al. 1977). Therefore, the composition of the flora may be regulated less by drought stress than by an ability to grow rapidly at high photosynthetic rates in response to chronic disturbance. Nitrogen fixation within the rhizosphere also may be an important adaptation that enables some of these species to colonize hostile new substrates.

Seedling establishment on new tephra deposits appears to require the development of a long taproot quickly because unaltered tephra, unlike the more mature soils below, holds little moisture or

FIG. 5.5   On the west slope, vegetation occurs well up the cone and appears to become sparse primarily in response to slope instability. The large, nearly bare area in the foreground was deglaciated during the May 18 eruption. Ridges support extensive, dense vegetation and may provide a source for revegetation of these glacial valleys. (Photo taken September 10, 1980.)

nutrients. Forbs are thus likely to have an advantage over most shallow-rooted graminoid species in colonizing barren volcanic substrates.

GERMINATION TRIALS

Table 5.4 summarizes the results of trials to germinate ten species in four substrates collected from the volcano. *Hieracium, Lupinus, Antennaria,* and *Potentilla* displayed high germination in coarse mudflow material and generally lower germination in upper mudflow and pyroclastic material. It was surprising to find that any seeds germinated in the claylike pyroclastic material. These preliminary results implied that the seeds of several species can germinate in pyroclastic material and fine muds. Soil toxicity may not be a major problem, but texture and limited nitrogen and other nutrients may be. Other species were unable to germinate in the densely packed pyroclastic materials, suggesting that reduced soil oxygen may prevent invading seeds from germinating except under unusually favorable conditions. Additional species have since undergone germination and growth trials to answer some questions raised by this exploratory study.

DISCUSSION AND CONCLUSIONS

Vegetation recovery following any major catastrophe results from two fundamental mechanisms: (1) the resurgence of surviving individuals, and (2) the establishment of colonists. Thus far, the first mechanism has vastly outweighed the second in importance in both the subalpine zone and other zones (J. F. Franklin, D. B. Zobel, J. A. Antos, J. E. Means, and M. A. Hemstrom, individual personal communications 1980, 1981).

Ridges and slopes above glacier fed streams received varying degrees of tephra deposition, but because of the protection of snow, herbaceous vegetation was little affected. Ecosystem function will likely return to normal within 2–4 yr in areas where the soil and plant cover were not severely modified. Where melting glaciers and ice fields deposited thin layers of mud on ridges, most of the plants initially survived. Few plants, however, were able to emerge through the often cementlike dried mud that caked the surface, and mortality has been high. Erosion has removed mud from the sites under investigation, revealing large barren areas that will require colonization by disseminules before the vegetation can return to a normal state.

In many places mudflows scoured away most of the existing vegetation. The availability of species adapted to the peculiar cold, sterile conditions found beneath mountain glaciers is in question. We hope to determine whether atypical patterns of recolonization or novel communities will develop in the absence of the normal pioneer species. Where thick layers of mud were deposited, vegetation will be established by two mechanisms. By the end of 1980, the vegetation of such flows consisted entirely of widely scattered plants that had survived transport in the flow and had resprouted from rootstocks. By a wide margin, *Lupinus lepidus* was the most common survivor, although large areas had to be searched to encounter even one individual. Seeds from surrounding areas will most likely become established in the coming years to augment seeds produced by the survivors. Establishment of vegetation on stabilized mudflows will be slow, but relatively steady.

Finally, many square kilometers of subalpine vegetation were completely destroyed by pyroclastic flows. Soil (in any meaningful sense) is lacking, and sources for colonization are generally remote, suggesting that the pace of vegetation recovery will be slow. Abraham Plains, covered with a coarse pumice layer, and the pyroclastic zone below the lip of the crater may long defy successful seedling establishment.

Future studies on the subalpine slopes of Mount St. Helens will focus on mechanisms of soil development, early plant succession, the roles played by insects in the recovery process, the contrast between totally devastated sites and those in which some individual plants and animals survived, the nature of physiological adaptations to stresses imposed by tephra and mud deposits, the rates of ecosystem recovery under various conditions, and the integration of these and other studies. Examples of specific questions being posed include: (1) What factors control the development of available nitrogen in the new substrates? (2) Are all species capable of growth on raw mud and pyroclastic materials or do some species require conditions mitigated by pioneering vascular plants? and (3) What adaptive mechanisms contribute to the differences in survival observed between species buried by tephra and those buried by mud? Such coordinated studies will enhance our understanding of ecosystem responses to profound perturbations and will thus improve our ability to predict ecosystem development and to prescribe means to accelerate recovery and mitigate adverse environmental impacts.

ACKNOWLEDGMENTS

We thank Jerry F. Franklin, U.S. Forest Service, for facilitating access to the volcano and providing helicopter transport. Field studies were assisted by Ted Thomas, Peter Frenzen, Karen Bliss, and Nancy Weidman. Robin Reid helped identify plant specimens and Becca Hanson reviewed the manuscript, providing numerous useful suggestions. These studies were supported by Grant DEB 80-21460 from the National Science Foundation.

LITERATURE CITED

CRANDELL, D. R., and D. R. MULLINEAUX. 1978. Potential hazards from future eruptions of Mount St. Helens volcano, Washington. *U.S. Geol. Surv. Bull. 1383-C*, U.S. Govt. Printing Office, Washington, D.C.

DEL MORAL, R. 1983. Initial recovery of subalpine vegetation on Mount St. Helens, Washington. *Amer. Midl. Nat.* 109:72–80.

GAUCH, H. G., Jr. 1982. *Multivariate Analysis in Community Ecology.* Cambridge University Press, Cambridge, U.K.

GRIGGS, R. F. 1919. Scientific results of the Katmai expeditions of the National Geographic Society. I. The recovery of vegetation at Kodiak. *Ohio J. Sci.* 19:31–342.

HILL, M. O. 1979. *DECORANA: A FORTRAN Program for Detrended Correspondence Analysis and Reciprocal Averaging.* Cornell University, Ithaca, N.Y.

HILL, M. O., and H. G. GAUCH. 1980. Detrended correspondence analysis: An improved ordination technique. *Vegetatio* 42:47–58.

HITCHCOCK, C. L., and A. CRONQUIST. 1973. *Flora of the Pacific Northwest.* University of Washington Press, Seattle.

LAWRENCE, D. B. 1941. The floating island lava flow of Mt. St. Helens. *Mazama* 23:56–60.

———. 1954. Diagrammatic history of the northeast slope of Mt. St. Helens, Washington. *Mazama* 36:41–44.

MAEDA, T., H. TAKENAKA, and B. P. WARKENTIN. 1977. Physical properties of allophane soils. *Adv. Agronomy* 29:229–264.

ROSENFELD, C. L. 1980. Observations on the Mount St. Helens eruption. *Amer. Sci.* 68:494–509.

ST. JOHN, H. 1976. The flora of Mt. St. Helens. *The Mountaineer* 70:65–78.

TABLE 5.1

GENERAL DESCRIPTION, SPECIES COMPOSITION, CANOPY COVERAGE (C), AND FREQUENCY (F) OF 24 0.1-m² QUADRATS AT BUTTE CAMP, MOUNT ST. HELENS[a]

| | Tephra deposits | | | | Steep slopes | | Mudflows | |
| | Low elevation | | High elevation | | | | | |
	F	C	F	C	F	C	F	C
Number of plots	4		6		7		6	
Mean cover (%)	19.5		21.2		6.8		0.2	
Mean richness (N)	12.0		12.7		12.0		3.8	
Total richness (N)	19		21		26		14	
Mean diversity (H')	1.81		2.20		1.65		0.74	
Species								
Achillea millefolium	—	—	7	0.6	2	0.1	1	t[b]
Agoseris aurantiaca	—	—	4	0.3	1	t	3	t
Agrostis diegoensis	76	6.1	51	3.4	25	1.0	3	0.1
Aster ledophyllus	21	0.6	18	0.9	6	0.2	1	t
Carex pachystachya	2	0.3	7	0.6	5	0.5	—	—
Castilleja miniata	1	0.1	4	0.1	2	0.1	—	—
Danthonia intermedia	9	0.4	16	0.8	—	—	4	0.1
Eriogonum pyrolifolium	40	2.3	29	1.4	32	1.2	—	—
Fragaria virginiana	1	0.1	15	1.0	5	0.1	1	t
Juncus parryi	13	0.5	33	1.5	11	0.5		

TABLE 5.1

GENERAL DESCRIPTION, SPECIES COMPOSITION, CANOPY COVERAGE (C), AND FREQUENCY (F) OF 24 0.1-m^2 QUADRATS AT BUTTE CAMP, MOUNT ST. HELENS[a](continued)

	F	C	F	C	F	C	F	C
Lomatium martindalei	58	3.4	17	0.4	1	t	—	—
Luetkea pectinata	—	—	5	0.2	1	t	1	t
Lupinus latifolius	2	t	—	—	—	—	—	—
Lupinus lepidus	54	4.9	46	5.6	17	1.0	7	0.2
Penstemon cardwellii	—	—	—	—	8	0.4	2	t
Phlox diffusa	—	—	16	1.5	7	0.5	1	t
Polygonum newberryi	11	0.3	15	1.6	8	0.5	4	t
Sitanion jubatum	6	0.3	6	0.3	3	0.2	—	—
Spraguea umbellata	10	0.2	1	t	5	t	—	—
Stipa occidentalis	5	0.5	3	0.1	2	0.2	—	—

[a]Nomenclature from Hitchcock and Cronquist 1973. Data collected Sept. 8–11, 1980.
b t = trace cover.

TABLE 5.2

GENERAL DESCRIPTION, SPECIES COMPOSITION, CANOPY COVERAGE (C), AND FREQUENCY (F) AT PINE CREEK STUDY SITE, MOUNT ST. HELENS

	Thin tephra		Thin mudflow		Thick mudflow	
Number of plots	4		4		5	
Mean cover (%)	20.4		1.1		0.3	
Mean richness (N)	10.8		5.0		2.6	
Total richness (N)	17		10		8	
Mean diversity (H')	1.07		1.19		0.9	
Species	F	C	F	C	F	C
Achillea millefolium	5	0.4	2	t[a]	—	—
Agoseris aurantiaca	4	0.7	1	t	—	—
Agrostis diegoensis	14	0.5	18	0.7	1	t
Aster ledophyllus	12	0.6	4	0.4	1	t
Carex spectabilis	34	5.6	—	—	—	—
Castilleja miniata	1	t	—	—	—	—
Danthonia intermedia	5	0.1	—	—	1	t
Eriogonum pyrolifolium	—	—	—	—	5	0.1
Fragaria virginiana	4	0.2	—	—	—	—
Juncus parryi	—	—	3	0.1	1	t
Lomatium martindalei	5	0.1	—	—	—	—
Luetkea pectinata	3	0.1	2	t	2	t
Lupinus latifolius	50	12.9	1	0.1	—	—
Lupinus lepidus	—	—	1	t	2	t
Penstemon cardwellii	—	—	—	—	2	t
Polygonum newberryi	9	0.3	7	0.1	—	—

[a]t = trace cover.

TABLE 5.3

GENERAL DESCRIPTION, SPECIES COMPOSITION, CANOPY COVERAGE (C),
AND FREQUENCY (F) AT RIDGES ABOVE THE
SOUTH FORK TOUTLE, MOUNT ST. HELENS

	Dry ridge		Mesic ridge	
Number of plots	2		2	
Mean cover (%)	21.5		52.5	
Mean richness (*N*)	12.0		19.0	
Total richness (*N*)	18		25	
Mean diversity (*H'*)	1.44		1.98	
Species	F	C	F	C
Achillea millefolium	35	4.1	82	13
Agoseris aurantiaca	—	—	13	0.9
Agrostis diegoensis	50	7.4	13	0.6
Aster ledophyllus	2	t[a]	—	—
Bromus carinatus	—	—	15	3.0
Carex spectabilis	—	—	27	2.0
Carex pachystachya	8	0.2	—	—
Castilleja miniata	4	t	2	t
Eriogonum pyrolifolium	4	0.4	1	t
Fragaria virginiana	25	2.8	44	3.0
Lomatium martindalei	2	0.1	19	0.6
Luetkea pectinata	2	t	—	—
Lupinus latifolius	9	0.2	44	18.8
Lupinus lepidus	—	—	6	0.3
Penstemon confertus	33	5.1	21	1.3
Phlox diffusa	—	—	27	1.8
Polygonum newberryi	4	0.1	—	—
Sitanion jubatum	2	t	16	1.0
Spraguea umbellata	4	t	—	—

[a]t = trace cover.

TABLE 5.4
Preliminary Germination Results in Four Substrates from Mount St. Helens

Species	Pyroclastic powder		Butte Camp mud		Butte Camp tephra		North Fork Toutle mud	
	T[a]	%[b]	T	%	T	%	T	%
Eriogonum pyrolifolium	3	12	—	0	4	25	5	12
Hieracium gracile	6	30	3	55	4	38	3	95
Lupinus lepidus	3	20	3	45	3	40	3	60
Luetkea pectinata	—	0	—	0	5	2	5	8
Luzula piperi	—	0	—	0	8	3	11	8
Penstemon cardwellii	14	3	—	0	14	5	11	5
Antennaria lanata[c]	3	40	3	40	3	72	3	65
Polygonum bistortoides[c]	—	0	—	0	13	20	13	5
Potentilla flabellifolia[c]	5	25	—	0	5	65	4	55
Valeriana sitchensis[c]	7	25	8	10	7	10	5	10

[a]T = number of days to first emerge.

[b]% = percentage of seedlings emerging after 30 days.

[c]Species collected in the Tatoosh Range and not encountered in the dry habitats of the study area.

6

First-Year Recovery of Upland and Riparian Vegetation in the Devastated Area around Mount St. Helens

Arthur McKee, Joseph E. Means,
William H. Moir, and Jerry F. Franklin

ABSTRACT

At the end of the first growing season following the May 18, 1980, eruption of Mount St. Helens, no vascular plants were found in regions subjected to pyroclastic flows, and plant cover on the debris flow within the blowdown zone was estimated at $10^{-6}\%$. The greatest plant cover within the devastated area was found on sites clearcut before the eruption; for these sites, the mean was 3.8% and the maximum sampled value was 17.2%. Areas of blown-down forests without a snowpack at the time of eruption had a mean plant cover of only 0.2% (maximum value of 0.66%). The presence of a snowpack during the eruption greatly ameliorated the effects of the blast and the associated deposits of ash on understory plants. Riparian areas had the greatest species richness, probably because of more favorable microsites found along streams and created by their action. Favorable microsites were also critical for plant survival and regrowth in upland habitats. Virtually all the live plants were perennials that had sprouted from the preeruption soil and had penetrated the ash or had been protected by a snowpack.

INTRODUCTION

The violent eruption of Mount St. Helens on May 18, 1980, devastated 61,000 ha of forested terrain north of the volcano (USDA Forest Service 1981a). The features of that and subsequent eruptions are described in detail by Lipman and Mullineaux (1981). The tremendous landslide, the powerful lateral blast, subsequent mudflows, ashfalls, and pyroclastic flows all combined to create a variety of new habitats for revegetation (Fig. 6.1). Areal extents of these major habitats are estimated in Table 6.1.

The landslide resulting from the collapse of the north flank of Mount St. Helens on May 18 formed a debris flow that moved 24 km down the North Fork of the Toutle River. Primarily consisting of rubble from the mountain, it also included some topsoil and various plant parts such as rhizomes, roots, and stems that could possibly sprout under appropriate conditions. To the northwest, north, and northeast of the peak, forests as far away as 28 km were flattened by the powerful lateral blast, forming a blowdown zone. A scorch zone of standing dead trees that had been killed by heated gases formed a band of varying width around the blowdown zone. Within the blowdown

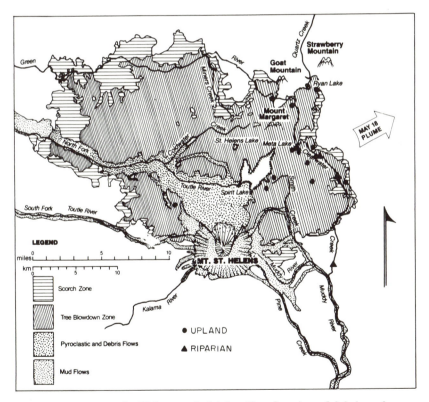

FIG. 6.1   Mount St. Helens and vicinity. Note location of debris and pyroclastic flows, tree blowdown, scorch zone, and mudflows.

zone, nearly all above-ground plant parts not buried in snow were killed on May 18. In the scorch zone, although mortality was extreme, some above-ground plant parts survived the eruption. Rapidly melting snow and ice and other processes created mudflows that filled stream channels and buried existing soils at depths of several meters. As in the debris flow, plant parts that could potentially sprout were incorporated into mudflows.

Ejecta from the blast and subsequent ashfalls formed deposits of tephra ranging from 10 to 60 cm deep or more. Pyroclastic flows associated with the explosive eruptions of ash raced down the north flank of the volcano, fanning out over the debris flow and areas subjected to earlier pyroclastic flows. The pyroclastic flows were totally devoid of plant parts that could reproduce vegetatively and thus formed a habitat for primary succession.

Prior to the May 18 eruption, the upland and riparian vegetation consisted of a productive conifer forest dominated by *Pseudotsuga menziesii, Abies amabilis, Abies procera, Tsuga heterophylla,* and *Thuja plicata.* The forests ranged in age from young stands in areas recently clearcut to old-growth forests over 500 yr in age (USDA Forest Service 1981b). The dominant trees in mature stands were >1 m in diameter at breast height and >50 m tall. Basal areas of such stands ranged from 60 to 110 m²/ha.

Logging over the last three decades had created an array of stands in various stages of secondary succession. After the eruption, these stands could be expected to have very different capacities for vegetative recovery than those of mature forests. Riparian vegetation also should respond differently from upland vegetation, which is subjected to less favorable moisture conditions.

Aerial reconnaissance prior to the study revealed that conspicuous differences existed in the first-year recovery of vegetation in the major habitats within the devastated area, which included the debris flow, mudflows, pyroclastic flows, clearcuts, blowdown, and standing dead forests. Moreover, the presence of a snowpack at the time of the May 18 eruption had apparently enabled some low shrubs and seedlings to survive. Riparian vegetation also appeared to be recovering more rapidly than the upland vegetation.

In light of these preliminary observations, the primary objectives of this study were as follows: (1) to document first-year patterns of revegetation in the major habitats created within the devastated area; (2) to compare vegetative recovery in forested areas clearcut prior to the eruption, in blown-down forests, and in standing dead forests; (3) to investigate the effect of snowpack in the blown-down forests on plant recovery; (4) to compare recovery of riparian vegetation on sites in the devastated area with that on sites receiving only ashfall; and (5) to establish a network of permanent plots for the study of vegetative recovery in the future.

METHODS

Several sampling methods were required to deal effectively with the range of spatial variation in the major habitats. Because of their barren nature, debris and pyroclastic flows were sampled by establishing transect lines and noting the presence of a plant species within 5 m of the transect. Areas subjected to the pyroclastic flows were sampled at three sites ranging from 1020 to 1200 m in elevation. At each site, three transects 250–300 m in length were sampled. A 2750-m transect was established at 800 m elevation on the debris flow in the valley of the North Fork of the Toutle River.

Clearcuts and blown-down and scorched forests were sampled by transects consisting of five 250-m² circular plots spaced 25 m apart. These transects were normally installed as adjacent pairs (occasionally triplets) to allow comparisons among clearcuts and blown-down and scorched stands. The percentage of cover for each species was estimated in each 250-m² circular plot. The nature of the tephra was described and its depth measured at least once in every plot. In every plot the coverage of woody debris was measured by an intercept line along a diameter placed at right angles to the direction of blowdown. A total of 35 transects were established in 1980: 13 in clearcuts, 13 in blown-down forests without a snowpack during the eruption, 6 in blown-down forests with a snowpack during the eruption, and 3 in scorched forests. Sites that had a snowpack during the eruption were readily recognizable because the tephra had slumped and cracked distinctively as the snow had melted from beneath.

The riparian vegetation along streams was sampled in 10-m-wide belt transects oriented perpendicularly to the main streamflow. The transects spanned the active, border, and outer riparian zones as defined by Campbell and Franklin (1979). The active zone included that portion of the riparian habitat subject to normal annual streamflow. The border zone included any area from which litter could reach the stream after some delay. It included, but was not restricted to, areas that flood during unusually high flows. The outer zone included the area within the influence of the higher water table near a stream, as evidenced by topography, vegetation, or both.

Five transects were established for riparian vegetation in 1980: three in the blast zone and two in nearby areas with just ashfall. Along the ashfall transects, plant cover was sampled in 0.2 × 0.5 m microplots spaced 1 m apart on 10-m-long subtransects perpendicular to the main transect and spaced 5 m apart. Species cover was assigned to one of six classes at each microplot, and the midpoint of the range of the cover class was used during the data reduction (Daubenmire 1959).

Vegetation was so sparse in the devastated area that cover for a species was estimated on the 5 × 5 m subplots formed by the main transect and subtransects. Because microsites apparently played a key role in the recovery patterns of riparian vegetation in the devastated area, the percentage of each 5 × 5 m subplot occupied by different microsites was estimated. These microsites were later grouped into six types: (1) steep banks and streambanks, (2) areas beneath elevated logs, (3) rootwads and tops of logs, (4) sand or gravel bars and overflow channels, (5) active channels, and (6) intact tephra. Tephra deposits were always reduced in the first three types of microsites because of the site steepness or the shelter provided by a fallen log. Tephra was often <5 cm deep on these microsites. The intact tephra microsites had undergone little or no disturbance or removal of the tephra since deposition.

The riparian vegetation around Meta Lake, located 13 km northeast of the volcano at 1080 m elevation, was sampled by five transect lines. Three of these were in the emergent wetland zone, and two were in the surrounding scrub shrub zone (Cowardin et al. 1979). Each transect consisted of 30 0.2 × 0.5 m microplots spaced 1 m apart. Plant species and seedbed substrates were assigned to one of six cover classes, and the midpoints of the ranges of the cover classes were used during data reduction (Daubenmire 1959).

Reconnaissance and sampling were conducted over a 2-week period in mid-September 1980. Helicopters were required to travel into the devastated area, and safety regulations limited the distance that sampling crews could travel from the landing sites. Because of difficult access, most sampling was confined to the northeastern portion of the devastated area (Fig. 6.1).

RESULTS

UPLAND VEGETATION

No vascular plants were found along the transects in areas subjected to pyroclastic flows, nor were any revealed on these barren surfaces by extended reconnaissance. The debris flows were almost as devoid of vascular plants as the pyroclastic flows. In September 1980, total plant cover on the 2750-m transect was estimated at $10^{-6}\%$; it consisted of only two small plants sprouted from rootstocks: a fireweed, *Epilobium angustifolium*, and a lady fern, *Athyrium filix-femina*.

Total plant cover was very low at all sites in clearcuts, blown-down forests without snowpack on May 18, and scorched forests (Table 6.2). The maximum value of 17.2%, obtained in a clearcut stand, was several times the maximum sampled in blown-down forests without snowpack or scorched forests. None of the sites sampled in the latter two habitats had even 1% total live plant cover. When two-way comparisons of plant cover and number of taxa per transect were made with a paired *t*-test (Snedecor and Cochran 1967), clearcuts had significantly more plant cover than did blown-down forests without snow ($p = 0.05$). They also possessed 10 times more plant cover than did the scorched forests, although the difference was not significant (probably because of the limited sampling of scorched forests).

Species richness in the three upland habitats was similar to cover at the end of the first growing season (Table 6.2); the number of taxa per transect was low in all habitats sampled. The mean number of taxa for clearcuts was significantly larger than that for blown-down forests without snowpack ($p = 0.10$) or scorched forests ($p = 0.01$), being about twice as large.

Almost all the plant species encountered on the transects had sprouted from plant parts located beneath the preeruption soil surface, but there were two types of exceptions. The most common was grass seedlings established from a seeding program by the USDA Soil Conservation Service in the late summer of 1980 (Stroh and Oyler 1981). The other exception was found in scorched forests near the margin of the devastated area, where the hot blast cloud had apparently cooled sufficiently so that occasional stems of deciduous shrubs survived and sprouted epicormically.

The presence of a snowpack at the time of the May 18 eruption ameliorated the effects of the blast on the understory vegetation (Table 6.3). In stands located between 1100 and 1340 m elevation, blown-down forests with snowpack had significantly greater plant cover than those without snowpack ($p = 0.05$). Although there was no significant difference in species richness between the two habitats, the maximum species richness ($n = 21$) among all the upland transects occurred in a blown-down forest with snowpack. Unfortunately, no clearcuts or scorched forests with snowpacks could be sampled for comparison.

The frequency of the most common taxa found on the transects in the upland habitats is shown in Table 6.4. The variation in plant species in the various habitats partially explains the differences observed in first-year vegetative recovery. The most common species in the clearcut areas were early successional plants that sprout vigorously and are well adapted to full sunlight (e.g., fireweed, *Epilobium angustifolium*; pearly everlasting, *Anaphalis margaratacea*; and thimbleberry, *Rubus parviflorus*). In contrast, common shrubs and saplings found at higher elevations (such as huckleberries, *Vaccinium* spp.; silver fir, *Abies amabilis*; and mountain hemlock, *Tsuga mertensiana*) dominated the blown-down forests protected by snowpack. In addition to the thermal insulation provided by the snowpack, the mechanical processes of cracking and slumping of the tephra as the snow melted from beneath provided thin spots and avenues whereby plants could more easily reach the surface. The sites with snowpacks also tended to be on the north-facing slopes in the lee of the blast.

The most common species encountered on transects in blown-down forests without snowpack were low shrubs. These had sprouted from plant parts located beneath the old soil surface. Most of the individuals were depauperate, in marked contrast to the vigorous early successional plants found in clearcuts. While the distribution of plants in microsites was not quantified, the plants in blown-down stands without snowpack were most commonly found on steeply angled root wads or under fallen tree boles at sites where the tephra had not accumulated as deeply as on open, level ground.

The relationship between plant cover and tephra depth in clearcut areas is complex because of the interactions of many factors, including the intensity of blast, temperature of deposits, composition of the tephra, and time since clearcutting occurred. The correlation of total plant cover with depth of tephra is poor (Fig. 6.2). The highest plant cover occurred at a site covered with slightly more than 20 cm of tephra. Some of the lowest values, however, were found at sites with this same depth of tephra. The site with the highest cover was deeply rilled, and plant recovery was greater in those microsites with thinner tephra. Erosion within the devastated area will clearly play a major role in vegetative recovery.

Earlier reconnaissance flights suggested a gradient of increasing plant cover as the distance from the volcano increased. This was not unexpected, because the intensity of the blast diminished away from the crater. The tephra depth also was reduced as the distance from the volcano increased. Except for one site, total plant cover in clearcuts increased with distance from the volcano (Fig. 6.3). At distances of <15 km from the volcano (where tephra depths were >20 cm), the cover declined to almost zero. Without exposure of the soil by erosion, plants had a difficult time penetrating tephra depths of >20 cm (Fig. 6.2).

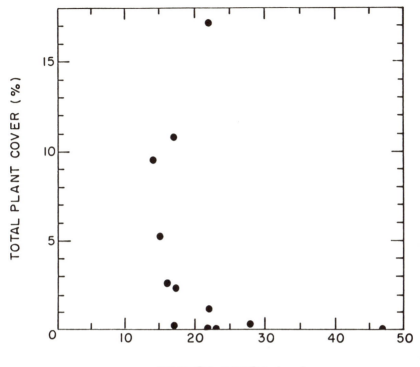

FIG. 6.2   Plant cover in September 1980 in areas clearcut prior to the May 18 eruption versus depth of tephra. Tephra depth determined by a minimum of five soil pits per site.

If tephra depth and distance from the volcano are held constant, the major upland habitats can be ordered from the greatest total plant cover to the least as follows: clearcut, blown-down forest with snowpack, scorched forest, and blown-down forest without snowpack. The ranking for species richness from greatest to least is: blown-down forest with snowpack, clearcut, scorched forest, and blown-down forest without snowpack. The eruption drastically reduced plant cover and species richness within the devastated area. Indeed, the values on the disturbed sites often approached or equaled zero. Undisturbed forested sites in the region would normally have plant covers of 100% or more, and species richness on plots of similar size would be five to ten times greater (Dyrness 1973, Dyrness et al. 1974).

RIPARIAN VEGETATION

Despite the apparent greenness of riparian areas from the air, sampling indicated that vegetative recovery was occurring there at about the same rate as on upland sites or perhaps even more slowly. Recovery during the first growing season depended on microsite conditions, and (as with upland sites) all riparian individuals in the devastated area resulted from vegetative reproduction from below-ground plant parts that survived the blast.

Total plant cover in riparian areas of the devastated area was very low, especially when compared to riparian transects in the nearby ashfall area (Table 6.5). The plant cover in the border and outer riparian zones of the ashfall area was within the range expected in the Pacific Northwest in undisturbed areas (Henderson 1978, Campbell and Franklin 1979). Values for total plant cover were low in the active zone for both the devastated and the ashfall regions, probably because of scouring by rapidly flowing streams heavily loaded with ash and pumice.

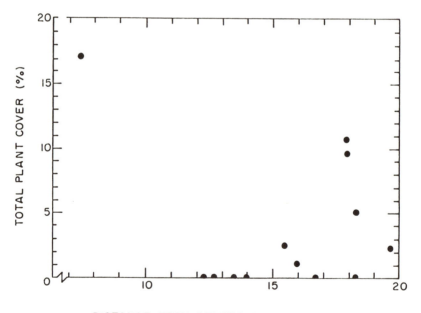

FIG. 6.3   Plant cover in September 1980 in areas clearcut prior to the May 18 eruption versus distance from the center of the volcano.

Little difference in species composition existed between the riparian habitats in the ashfall and in the devastated areas (Table 6.6). The species found in riparian areas were typical of riparian sites west of the Cascade range (Henderson 1978, Campbell and Franklin 1979).

After adjusting for varying plot sizes, species richness was greater in the ashfall area than in the devastated area (Table 6.5). This pattern was consistent with earlier preliminary observations. Some similarity occurred in the number of species encountered in the border zones of the two habitats, possibly because the overlying tephra was removed by high streamflows with limited scouring during storms. Such action would permit sprouting through the thinner layer of tephra and consequent growth without major scouring damage.

Previous reconnaissance suggested that microsite differences were important in riparian habitats in the devastated area. Accordingly, the riparian transects were stratified by microsite and cover values normalized for microsite area (Table 6.7). Microsites with relatively thin tephra deposits, such as streambanks and steeply tilted root wads, tended to have the highest cover values within a transect. The lowest values were found on intact or very slightly reworked tephra. The microsites with the most plant cover comprised only about 5% of the total area sampled by transects. These islands of green in an otherwise gray world will undoubtedly be an important source of propagules for recolonization of the riparian habitats in the devastated area.

Total plant cover and number of taxa per transect around Meta Lake are shown in Table 6.8. The highest cover in this shoreline area was found in the emergent zone, but slightly more taxa occurred in the scrub shrub zone, which is located at a slightly more elevated position around the lake. Both cover and species richness, however, were low in all the zones. The dominant species found in the three zones around Meta Lake are shown in Table 6.9. The species composition suggests the presence of a snowpack on the shore of the lake at the time of the blast. Only on blowdown sites with a snowpack were *Vaccinium* spp., *Menziesia ferruginea*, and especially *Abies amabilis* found with coverages as high as around Meta Lake.

The distribution of plant species around Meta Lake also suggested a possible interaction between snowpack and tephra deposition (Table 6.10). The north shore (which faces south) had a relatively gentle slope; consequently, little tephra washed into the lake. Also, this shore had either no snowpack or just a thin one at the time of the blast. The emergent zone species, *Carex aquatilis*, was protected from the blast by the water and was doing well. Vegetation in the scrub shrub zone not protected by a snowpack was impoverished, unlike that on the south shore. The north-facing south shore had a snowpack that offered some protection to plants of the scrub shrub zone. The south shore was steep, however, and an alluvial fan of tephra covered much of the emergent zone. The low cover values in the emergent zone were possibly a result of relocated tephra burying the emergent plants.

More streamside than lakeside sites were examined by both sampling and preliminary observation. It should be emphasized that the small number and restricted distribution of sampled sites precluded statistical testing of hypotheses concerning riparian recovery. The patterns observed, however, appeared to be consistent with more extensive observations at other sites.

DISCUSSION

While the overall impression of the devastated area in mid-September, 1980, was that its designated name was appropriate, vegetative recovery had begun on all of the major habitats except those areas subjected to pyroclastic flows. The recovery during the first growing season had several interesting characteristics:

1. Virtually all species of vascular plants present were perennials that had sprouted from below the soil surface and had penetrated the ash or were protected by snowpack.

2. Within the devastated zone, the presence of a snowpack at the time of the eruption was an important factor in moderating the effects of the blast.

3. Total plant cover was greatest in clearcut sites because of the ability of early successional plant species to reproduce vegetatively.

4. Total plant cover was minimal in blown-down forests without snowpack because of the apparent inability of forest understory species to resprout vigorously.

5. Species diversity was greatest in riparian habitats; this was a function of the diversity of microsites created by the action of the streams.

6. Within riparian habitats, the active zone was extremely low in both plant cover and species richness.

7. The emergent zone had greater plant cover than the scrub shrub zones around lakes, but the latter had greater species richness.

8. Microsite conditions were key factors in survival and reestablishment in both riparian and upland habitats.

9. Strong environmental and vegetational gradients existed within the devastated area, with both elevation and distance from the mountain.

The future pattern of revegetation will be a function of several factors, including the influx of plant propagules, rates of vegetative reproduction, erosional processes on upland and riparian sites, logging and reforestation activities, and of course, future eruptions of the mountain. The soil that lies buried beneath the tephra also will play a key role in revegetation. This soil once supported productive forests and remains intact over most of the area. Virtually all of the vegetation appearing during the first growing season sprouted from, or was rooted in, this buried soil. These plants will serve as important sources of propagules.

The tephra is a poor medium for plant establishment and growth. Samples of the tephra taken within 30 km of the volcano were very low in nitrogen (0.018%) and phosphorus (4.81 ppm) and had a poor cation exchange capacity (2.86 meq/100 g) (unpublished data, G. Klock, USDA Forest Service, Wenatchee, Washington). The trial seedings by the USDA Soil Conservation Service were most successful when grass and legume species were able to reach the buried soil (Stroh and Oyler 1981). Those plants that were rooted only in the tephra showed severe nutrient deficiencies.

The recovery of this devastated area will most likely be rapid relative to many other volcanic eruptions because the soil buried beneath the tephra is within the rooting distance of many plants. Lava flows and thick tephra deposits characteristic of other eruptions slow revegetation and couple it to soil formation processes. Smathers and Mueller-Dombois (1974) provide a good review of posteruption revegetation.

Removal of the overlying tephra by erosion will hasten upland revegetation by allowing sprouts from buried vegetation to reach the surface more easily and by reducing the distance a developing root must grow before entering the nutritionally rich soil. Rapid revegetation after exposure of the buried soil has been observed at Katmai National Monument (Griggs 1919) and Paricutin Volcano (Eggler 1948).

Recovery of the riparian vegetation will possibly be adversely affected by the erosional processes that enhance hillslope revegetation. The erosion of tephra into the riparian zone with subsequent reworking by fluvial processes could dramatically slow the recovery of the riparian vegetation. Regrowth may be buried by aggrading streams or severely scoured by the higher flows. Under heavy loads of sediment, stream channels will be less stable and the rate of new channel formation will be increased, with consequent undercutting and removal of streambank vegetation. For the next few years, the woody debris in and near stream channels will probably have a greater influence on channel morphology than will the recovering vegetation. Removal of the coarse woody debris by salvage logging will further destabilize the system and may have a profound effect on geomorphic processes and recovery rates of the riparian vegetation.

Further study of the devastated area should result in a better understanding of ecosystem recovery, particularly as it pertains to the landscape of the Cascade range. From Mount Rainier to Mount Lassen, the Cascade range has been formed by relatively recent volcanism (McKee 1972, Crandell et al. 1975, Heusser and Heusser 1980, Hoblitt et al. 1980). Thus the 1980 eruption of Mount St. Helens is not unique. Past volcanic activity has probably determined the current pattern of vegetation in the Cascades in such diverse ways as depositing tephra of different chemistries and depths and reducing or eliminating plant species in heavily impacted areas. Detailed study of the revegetation in the devastated area around Mount St. Helens should contribute to an understanding of the relative importance of various environmental factors on the region's forest and riparian ecosystems.

ACKNOWLEDGMENTS

Several people contributed to this paper. Miles Hemstrom collaborated in the initial sampling design for upland vegetation and led a field crew. Diane Mitchell, Sarah Greene, and Robert Frenkel assisted in riparian sampling. Fred Bierlmaier, Judy Alaback, Peter Frenzen, Glenn Hawk, and Robin Graham were among the many people of the field crew who worked for long periods and contributed many ideas. This study was supported by the Pacific Northwest Forest and Range Experiment Station of the USDA Forest Service, by Oregon State University, and by National Science Foundation Grant DEB 8024471 to Oregon State University. This is Paper 1648 of the Forest Research Laboratory, Oregon State University.

LITERATURE CITED

CAMPBELL, A. G., and J. F. FRANKLIN. 1979. Riparian vegetation in Oregon's Western Cascade Mountains: Composition, biomass, and autumn phenology. *Coniferous Forest Biome Bull. 14*, 90 pp. U.S./International Biological Program, Univ. of Washington, Seattle.

COWARDIN, L. M., V. CARTER, F. C. GOLET, and E. T. LaROE. 1979. Classification of wetlands and deepwater habitats of the United States. *FWS/OBS-79/31*, 103 pp. Office of Biological Services, Fish and Wildlife Service, U.S. Department of the Interior, Washington, D.C.

CRANDELL, D. R., D. R. MULLINEAUX, and M. RUBIN. 1975. Mt. St. Helens volcano: Recent and future behavior. *Science* 187:438–441.

DAUBENMIRE, R. F. 1959. Canopy coverage method of vegetation analysis. *Northwest Sci.* 33(1):43–64.

DECKER, R., and B. DECKER. 1981. The eruptions of Mount St. Helens. *Sci. Am.* 244 (3):68–80.

DYRNESS, C. T. 1973. Early stages of plant succession following logging and burning in the western Cascades of Oregon. *Ecology* 54(1):57–69.

DYRNESS, C. T., J. F. FRANKLIN, and W. H. MOIR. 1974. A preliminary classification of forest communities in the central portion of the western Cascades in Oregon. *Coniferous Forest Biome Bull. 4*, 123 pp. Univ. of Washington, College of Forest Resources, Seattle.

EGGLER, W. A. 1948. Plant communities in the vicinity of the volcano El Paricutin, Mexico, after two and a half years of eruption. *Ecology* 29(4):415–438.

GRIGGS, R. F. 1919. The main character of the eruption as indicated by its effects on nearby vegetation. *Ohio J. Sci.* 19(3):173–209.

HENDERSON, J. A. 1978. Plant succession on the *Alnus rubra/Rubus spectabilis* habitat type in western Oregon. *Northwest Sci.* 52(3):156–167.

HEUSSER, C. J., and L. E. HEUSSER. 1980. Sequence of pumiceous tephra layers and the late Quaternary environmental record near Mount St. Helens. *Science* 210: 1007–1009.

HOBLITT, R. P., D. R. CRANDELL, and D. R. MULLINEAUX. 1980. Mount St. Helens eruptive behavior during the past 1,500 years. *Geology* 8:555–559.

LIPMAN, P. W., and D. R. MULLINEAUX. 1981. The nineteen eighty eruptions of Mount St. Helens, Washington. *U.S. Geol. Surv. Prof. Paper 1250*, 844 pp. U.S. Government Printing Office, Washington, D.C.

MCKEE, B. 1972. *Cascadia*. McGraw-Hill, New York. 394 pp.

ROSENFELD, C. L. 1980. Observations on the Mount St. Helens eruption. *Am. Scient.* 68(5):494–509.

SMATHERS, G. A., and D. MUELLER-DOMBOIS. 1974. Invasion and recovery of vegetation after a volcanic eruption in Hawaii. *U.S. Nat. Park Serv. Scient. Monogr. Ser.* (No. 5), 129 pp. Washington, D.C.

SNEDECOR, G. W., and W. G. COCHRAN. 1967. *Statistical Methods*. (6th ed.). Iowa State Univ. Press, Ames, Iowa. 593 pp.

STROH, J. R., and J. A. OYLER. 1981. *Assessment of Grass–Legume Seedings in the Mt. St. Helens Blast Area and the Lower Toutle River Mud Flow, May 1981*. USDA Soil Conservation Service, Spokane, Washington.

USDA FOREST SERVICE. 1981a. *Draft Environmental Impact Statement—Mount St. Helens Land Management Plan*. Gifford Pinchot National Forest, Vancouver, Washington. 162 pp. plus 24 maps.

———. 1981b. *Preliminary Plant Association and Management Guide for the Pacific Silver Fir Zone of the Gifford Pinchot National Forest*. Pacific Northwest Region, Portland, Oregon, 123 pp.

U.S. GEOLOGICAL SURVEY. 1981. Mt. St. Helens and vicinity (map). Available from Gifford Pinchot National Forest, Pacific Northwest Region, USDA Forest Service, Vancouver, Washington.

TABLE 6.1

AREAL ESTIMATES OF HABITATS IN THE DEVASTATED AREA AROUND
MOUNT ST. HELENS, SEPTEMBER, 1980

Habitat	Area (ha)	Source of estimate[a]
Blowdown zone	35,000	2
Blown-down forest	21,000	1
Clearcuts	14,000	3
Scorch zone	11,600	2
Scorched forest	9,600	2
Clearcuts	2,000	3
Pyroclastic flow	1,000	5
Debris flow	7,100	4
Mudflows	5,600	2
Crater	700	2
Total	61,000	1 and 2

[a]Source of estimates: 1. USDA Forest Service 1981a; 2. estimated from Post Eruption Conditions Map in USDA Forest Service 1981b; 3. computed as the difference between values given in the above two references; 4. estimated from map in U.S. Geological Survey et al. 1981; 5. estimated from map in unpublished manuscript by N. C. Banks and R. P. Hoblitt, 1981, U.S. Geological Survey, Denver, Colorado. Only the area of the mudflows within the boundaries of the Post Eruption Conditions Map is given.

TABLE 6.2

MEAN PLANT COVER AND NUMBER OF TAXA PER TRANSECT IN
CLEARCUTS, BLOWN-DOWN FOREST WITHOUT SNOW,
AND SCORCHED FOREST IN SEPTEMBER, 1980

Habitat	Number of transects	Cover (%)		Number of taxa per transect	
		Mean	Range	Mean	Range
Clearcut	13	3.8[a]	0.0–17.2	6.2[b,c]	0–15
Blown-down forests without snow	13	0.2[a]	0.0–0.66	3.7[c]	1–8
Scorched forests	3	0.4	0.2–0.5	4.7[b]	3–7

[a]Significantly different at 0.05 according to a paired t-test.
[b]Significantly different at 0.10 according to a paired t-test.
[c]Significantly different at 0.01 according to a paired t-test.

TABLE 6.3

MEAN PLANT COVER AND NUMBER OF TAXA PER TRANSECT IN
SEPTEMBER, 1980, IN BLOWN-DOWN FORESTS WITH AND WITHOUT
SNOWPACK ON MAY 18, 1980

Habitat	Number of transects	Cover (%)		Number of taxa per transect	
		Mean	Range	Mean	Range
Blown-down forests without snow	5	0.06[a]	0.04–0.08	2.6	1–4
Blown-down forests with snow	6	3.3[a]	0.2–10.1	8.0	2–21

[a]Significantly different at 0.05 according to a paired t-test.

TABLE 6.4

FREQUENCY OF IMPORTANT TAXA IN CLEARCUTS, BLOWN-DOWN FORESTS WITH AND WITHOUT SNOWPACK ON MAY 18, 1980, AND SCORCHED FOREST[a]

Taxa	Frequency (%)			
	Clearcuts	Blown-down forests without snow	Blown-down forests with snow	Scorched forests
Abies amabilis (Dougl.) Forbes	0	9	40	0
Acer circinatum Pursh	0	0	0	47
Anaphalis margaratacea (L.) B. & H.	22	0	0	0
Cirsium spp.	14	0	0	0
Cornus canadensis L.	11	15	0	33
Epilobium angustifolium L.	58	0	3	0
Galtheria shallon Pursh	5	14	0	13
Hieracium albiflorum Hook.	11	0	0	7
Pteridium aquilinum (L.) Kuhn	11	0	0	0
Rubus lasiococus Gray	1	8	11	0
R. parviflorus Nutt.	32	0	0	0
R. spectabilis Pursh	11	0	0	0
Tsuga mertensiana (Bong.) Carr.	0	0	15	0
Vaccinium membranaceum Dougl.	3	14	22	20
V. ovalifolium Smith	6	32	32	47

[a]Clearcuts had 65 transects, blown-down forests without snow had 65 transects, blown-down forests with snow had 30 transects, and scorched forests had 15 transects. All taxa with frequency greater than 10% in September, 1980, are included.

TABLE 6.5

VASCULAR PLANTS IN ACTIVE, BORDER, AND OUTER RIPARIAN ZONES IN THE
DEVASTATED AND ASHFALL AREAS, SEPTEMBER, 1980

Area and transect number	Location	Herb and shrub cover (%)			Number of species			
		Active	Border	Outer	Active	Border	Outer	Whole transect
Devastated area								
345	Upper Clearwater Creek	t[a]	3	t	5	17	10	24
346	Upper Clearwater Creek	t	1	t	0	6	1	6
347	Middle Clearwater Creek	t	2	t	5	33	5	34
Ashfall area								
348	Lower Clearwater Creek	1	36	70	16	23	44	51[b]
350	Clear Creek	4	85	80	20	47[b]	38	60[b]

[a] t = trace cover (<0.5%).
[b] Sedges and grasses were identified generically and probably included several additional species.

TABLE 6.6

DOMINANT SHRUBS AND HERBS ON RIPARIAN TRANSECTS IN THE
DEVASTATED AND ASHFALL AREAS, SEPTEMBER, 1980[a]

Stratum and taxa	Cover (%)	
	Devastated area[b]	Ashfall area
Tall shrubs		
Acer circinatum Pursh	0.1	12.1
A. macrophyllum Pursh	t[b]	6.5
Alnus rubra Bong.	t	2.7
Cornus stolonifera Michx.	t	2.2
Low shrubs and herbs		
Festuca subulata Trin	0	1.7
Gymnocarpium dryopteris (L.) Newm.	t	1.4
Heracleum lanatum Michx.	0.1	0
Pteridium aquilinum (L.) Kuhn	0.3	0
Rubus parviflorus Nutt.	t	5.4
R. spectabilis Pursh	0.4	3.8
R. ursinus Cham. & Schlecht.	0	3.3
Stachys cooleyae Heller	t	1.9
Tiarella unifoliata Hook.	t	1.3
Tolmiea menziesii (Pursh) T. & G.	0	1.3

[a]Devastated area had 3 transects and ashfall area had 2 transects. All taxa with >1% cover in the ashfall area and >0.1% cover in the devastated area are included.

[b]t = trace cover (>0.1%).

TABLE 6.7

RIPARIAN PLANT COVER BY MICROSITE IN BLOWN-DOWN FORESTS
IN THE DEVASTATED AREA, SEPTEMBER, 1980[a]

Microsite	Total plant cover per transect number (%)			
	345	346	347	Mean
Streambanks	18.6	0.2	13.3	10.7
Areas under elevated logs	3.4	1.2	0	1.6
Rootwads and tops of logs	13.2	2.5	1.2	5.6
Bars and overflow channels	3.0	0	0.3	1.1
Intact tephra	1.7	0.3	0.2	0.7
Active channels	0	0	0	0

[a]The unweighted means for the three transects, which differ in area, are given.

TABLE 6.8
MEAN PLANT COVER AND NUMBER OF TAXA PER TRANSECT IN THREE ZONES AROUND META LAKE, SEPTEMBER, 1980

Zone	Cover (%)	Number of taxa per transect
Emergent	12.4	3.3
Scrub shrub	4.7	5.5
Forest	0.2	2.0

TABLE 6.9
MEAN COVER OF IMPORTANT TAXA IN THE SHORELINE TRANSECTS AROUND META LAKE, SEPTEMBER, 1980[a]

Zone[b] and taxa	Cover (%)
Emergent	
Carex aquatilis Wahl.	9.2
Mosses	1.1
Salix spp.	0.7
Vaccinium ovalifolium Smith	0.7
Scrub shrub	
Alnus sinuata (Regel) Rydb.	1.7
Menziesia ferruginia Smith	1.2
Trautvetteria caroliniensis (Walt.) Vail	0.7
Forest	
Abies amabilis (Dougl.) Forbes	0.2
Lichens	0.6

[a]All taxa with more than 0.5% cover in emergent and scrub shrub transects and all taxa in the forest transect are included.

[b]Emergent zone had 3 transects, scrub shrub zone had 2 transects, and forest zone had 1 transect.

TABLE 6.10
MEAN COVER OF SHRUBS AND HERBS IN THREE ZONES ON THE TRANSECTS AROUND THREE SIDES OF META LAKE, SEPTEMBER, 1980

Zone	Cover (%) (shrubs/herbs)		
	South shore	North shore	West shore
Emergent	2.2/0.15	2.3/28.6	0/0.1
Scrub shrub	5.9/2.0	0.5/0.5	—[a]
Forest	—	0/0	—

[a]— indicates no data available.

7

Vegetation Changes Induced by Mount St. Helens Eruptions on Previously Established Forest Plots

Miles A. Hemstrom and William H. Emmingham

ABSTRACT

Seventy-seven vegetation plots established before the 1980 eruptions of Mount St. Helens were relocated and remeasured in September, 1980. Impacts included burial by deep mudflows, destruction by the volcanic blast, and deposition of tephra in areas with and without tree canopy destruction. Plant species were variously affected by increasingly deep tephra deposits; many species, particularly herbs and trailing shrubs, were eliminated by tephra deposits >10 cm thick. The impact of the eruptions on plant communities was directly related to the presence or absence of snowpacks, the phenologic stage of the plants at the time of the eruption, and the life forms of the species involved. Difficulties were encountered in duplicating preeruption ocular cover estimates and in comparing plots that received different impacts. The plots provide permanent remeasurement points where successional recovery can be observed in the future.

INTRODUCTION

Ecologists rarely have the chance to study described plant communities that have been subsequently affected by major natural disturbances. Usually, they analyze succession as undefined plant communities recover from disturbance (Henry and Swan 1974, White 1979) or as documented communities recover from induced disturbances (Borman and Likens 1979). On the basis of such studies, ecologists determine the tendency of plant communities to resist disturbance-induced changes and the rapidity and degree to which they return to a previous state. Extrapolation of the results from these studies to natural disturbances is always clouded either by the "unnatural" character of the disturbance or by the original community composition being unknown. The 1980 eruptions of Mount St. Helens provide an

188

unusual opportunity to study succession on sites already established and sampled for a plant community classification (Brockway et al. 1982).

METHODS

During the summers of 1978 and 1979, field crews from the USDA Forest Service, Area VI Ecology Program, sampled vegetation between 640 and 1800 m elevation to the north, east, and south of Mount St. Helens. Circular sample plots 500 m^2 in area were established in relatively undisturbed, mature stands representing typical forest communities with the purpose of developing a plant association classification. The center and surrounding forest were photographed at each plot. Plot locations were recorded on maps and mosaics of aerial photographs. Crews recorded elevation, slope steepness, slope aspect, vascular plant species composition, a visual estimate of percentage of cover for each species, and soil information. Tree measurements included total age, total height, diameter at breast height, and diameter growth rate on at least one dominant canopy tree and stand basal area.

Eruptions of Mount St. Helens on and since May 18, 1980, disturbed over 100 of these sample plots located 2–100 km from the crater. Impacts included complete burial by mudflows, destruction of vegetation in the blast zone, and deposition of tephra as thick as 25 cm. In September 1980, one growing season following the major eruption of May 18, 77 of these plots were relocated for the following purposes: (1) to document changes in vascular plant species composition and dominance, (2) to establish a baseline for long-term remeasurement of plant community succession, (3) to identify changes in plant associations that could be used to revise a management-oriented classification of the Pacific silver fir zone, and (4) to examine the interaction between time of the tephra deposition and the degree of snowmelt or leaf-out of plants.

The 77 plots represent a gradient of disturbance intensity and community types (Table 7.1). About 60% of the plots were exactly relocated from maps, ground and aerial photographs taken prior to the eruption, and the presence of soil description pits dug during the original sampling. The remaining plots were placed as close as possible to the original location. All vascular plants present were listed, visual estimates were made of percentage of cover for each species, ash deposits were described by depths and layers, and any obvious morphological or pathological changes in leaf or stem characteristics were noted. Photographs were taken to duplicate as closely as possible the original stand photographs. All relocated plots were permanently marked with steel reinforcing bars and were carefully located on maps. Plots were ranked by disturbance intensity and tephra depth (Table 7.1). Preeruption and posteruption percentage of cover and presence of all vascular plant species in each plot were then compared.

RESULTS

Eight relocated plots were selected to illustrate the effects of differing disturbance intensities on the most common plant associations (Fig. 7.1). Plot 393 (Fig. 7.2) is located on mud and pyroclastic flow material 1 km south of Spirit Lake. The preeruption plant community

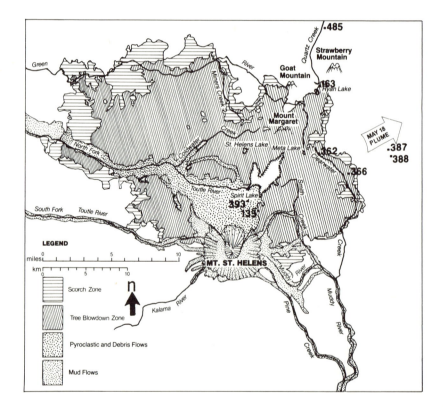

FIG. 7.1   Impacts from the May 18, 1980, eruption of Mount St. Helens and locations of the eight preeruption vegetation plots shown in Figures 7.2–7.9.

was dominated by *Pseudotsuga menziesii*, *Tsuga heterophylla*, and *Abies amabilis* with an understory of *Vaccinium* spp., *Menziesia ferruginea*, and a variety of moist site indicating herbs (Fig. 7.2; Table 7.2). Mudflows, blasts, and pyroclastic flows eliminated all vascular plant species. Total depth of deposits was >50 m. Succession is proceeding from a bare, nutrient-poor primary surface and most likely began with light-seeded invaders such as *Epilobium angustifolium* and *Circium* spp.

Plot 135 (Fig. 7.3), in the blast zone at the headwaters of Smith Creek, was not affected by mudflows. In addition to the initial blast, which killed and removed above-ground parts of all vascular plants, the plot was covered with 20 cm of tephra. Although this plot is on an east-facing slope 4 km from the crater, the tephra deposit here is only 20 cm deep—less than half that found on more protected slopes nearby. Some of the tephra may have been scoured away by the blast or eroded from the steep slope. The original plant community was composed of an overstory of *Abies procera*, *A. amabilis*, and *Tsuga mertensiana* and an understory dominated by *Vaccinium* spp., *Menziesia ferruginea*, *Tiarella trifoliata*, *Cornus canadensis*, *Clintonia uniflora*, and several other herbs (Table 7.2). No vascular plants were observed on the plot during remeasurement. Succession will be more rapid here than on Plot 393 because the original soil, which lies only 20 cm beneath the tephra surface, has been exposed by erosion of the tephra. Although burial for an entire year killed the rootstocks of many plants, those that were exposed by erosion during the 1980–81 winter and survived will rapidly recolonize the area. Similar nearby areas covered with thicker deposits may recover more slowly.

Relocated Plots 366, 163, and 362 are 14–22 km northeast of the crater on the edge of the blast zone (Fig. 7.1). The blast shattered the tree canopy in all three plots, and no living trees were present at remeasurement (Table 7.2). No living plants were recorded at Plot 362, which before the eruption was a species-poor community dominated by *Abies amabilis*, *Tsuga heterophylla*, *Pseudotsuga menziesii*, *Acer circinatum*, and *Chimaphila umbellata* (Fig. 7.4; Table 7.2). At Plot 163, near Ryan Lake over 20 km from the crater, leaves and branches were blasted from the trees, but only a few trunks were blown down (Fig. 7.5). The original community here was dominated by *Abies amabilis*, *Tsuga heterophylla*, *Vaccinium* spp., and a sparce herbaceous layer (Table 7.2). Trace amounts of several species survived the blast and the 13-cm-deep tephra deposit. At Plot 366 (Fig. 7.6), originally similar to Plot 362, several species survived the blast and 25 cm of tephra deposition, including *Acer circinatum*, *Berberis nervosa*, *Rubus lasiococcus*, *R. ursinus*, and *Cornus canadensis* (Table 7.2).

FIG. 7.2   Preeruption (A) and posteruption (B) photographs of Plot 393.

      Three plots illustrate the effects of tephra deposition on vegeta-
tion without canopy destruction. Plot 387 (Fig. 7.7), originally an *Abies
amabilis/Rhododendron albiflorum* association dominated by several
shrub species and a few herbs (Table 7.2), received >13 cm of tephra
deposition on top of a snowpack. Most of the shrub species were
drastically reduced. *Erythronium montanum*, the only herb present after
the eruption, was not listed before the eruption (probably a sampling
error). This species sprouts vigorously from bulbs in early spring,
often penetrating a few centimeters of snow, and in this special case

easily pushed through several centimeters of loose tephra.

Plot 388 is a typical example of the *Abies amabilis/Oplopanax horridum* association (Fig. 7.8). *Oplopanax horridum* appeared to be very resistant to tephra burial if tephra deposition occurred before leaf-out. Although 4 cm of tephra substantially reduced the cover of most shrubs and herbs, *O. horridum* cover remained constant (Table 7.2). Some of the more delicate species, especially *Rubus lasiococcus* and *Tiarella trifoliata*, were eliminated.

Plot 485 is a lower elevation stand dominated by *Pseudotsuga menziesii, Tsuga heterophylla, Acer circinatum, Gaultheria shallon,* and *Berberis nervosa* (Fig. 7.9). A 1-cm-thick tephra deposit only slightly reduced the cover of most shrubs and herbs (Table 7.2). Only fragile herbs and trailing shrubs, originally present in small amounts, declined substantially. A few species originally present in trace amounts were eliminated.

DISCUSSION

The thickness of tephra deposits after erosion the first winter is the most significant factor affecting changes in percentage of cover and species numbers, except in mudflow and blast areas. Plots not affected

FIG. 7.3   Preeruption (A) and posteruption (B) photographs of Plot 135.

by blast or mudflows were arranged in order of tephra depth and examined for the number of species before and after eruption as well as for the change in number of species with increasing tephra depth (Fig. 7.11). The mean change in number of species varied greatly and generally decreased with increasing tephra depth. The number increased, however, between 2 and 4 cm of tephra, apparently because of sampling error or population shifts on a few plots.

The life form of each species influences its response to burial by tephra. The effects of tephra on life forms, however, are compounded by the presence or absence of a snowpack during tephra deposition. Plots without a snowpack were examined for changes in percentage of cover of six species: (1) a fragile herb, (2) a robust herb, (3) a trailing shrub, (4) a low, erect shrub, (5) a tall, erect shrub, and (6) trees less than 3.5 m tall (Fig. 7.10). The effects of increasing tephra depth on these species are difficult to interpret except in a general way. The cover of all seven species declined when tephra depth exceeded 10 cm. Small trees appeared to be least affected. Species with stout stems or branches survived deep tephra deposits better than prostrate shrubs and fragile herbs. Impacts from <10 cm of tephra deposition varied

B

considerably because of the character of the deposit and the small sample sizes. Some of the substantial declines in cover that occurred at 2–3 cm of tephra appear to have resulted from tephra deposition during a rainstorm. The resulting deposit resembled a slurry of cement poured over the plant community, reducing cover more than a similar depth of dry tephra would have.

Although the number of species decreased at all depths from 1 to 25 cm of tephra deposition, there was no general pattern of increased number of species lost with depth. The number of species lost at 2.5 cm

FIG. 7.4   Preeruption (A) and posteruption (B) photographs of Plot 362.

FIG. 7.5   Preeruption (A) and posteruption (B) photographs of Plot 163.

FIG. 7.6   Preeruption (A) and posteruption (B) photographs of Plot 366.

was more than that lost at 17 cm. Only at 25 cm tephra depth did there seem to be a statistically relevant decrease in the number of species lost over those lost at 17 cm or less. Trends are apparent but anomalies occur as well. For example, at about 7.5 cm of tephra there was an increase in cover of most of the species compared to that found at 5 or 10 cm. Some of the variability may be due to the character of the tephra deposition at each site—its particle size distribution and layering. The variation in the interaction of plant community and tephra deposition precludes generalization about the effect of tephra depth on change in species cover except in one way: increasing tephra depth did eliminate more species.

Several sources of variation make it difficult to find statistically significant differences in the impacts of the various disturbances on plant communities. Different researchers performed visual estimates of percentage of cover and made species lists before and after the May 18, 1980, eruption. Although the investigators were experienced field workers, some differences in estimates were bound to occur. In addition, each plot experienced a unique disturbance with regard to (1) blast presence, intensity, and direction; (2) depth and consistency of

B

the tephra deposit; (3) snowpack presence or absence; and (4) subsequent tephra movement (deposition or erosion). The composition, physiognomy, and phenological stage (degree of leaf-out) of the plant community also influenced the impact a particular level of disturbance had on the plant community.

We intend to apply additional sampling methods in the future to ensure more repeatable measurements, including: (1) a systematically installed series of microplots (10 × 50 cm) along transects that can be exactly relocated within each plot, (2) carefully located and repeatable

FIG. 7.7   Preeruption (A) and posteruption (B) photographs of Plot 387.

FIG. 7.8   Preeruption (A) and posteruption (B) photographs of Plot 388.

FIG. 7.9   Preeruption (A) and posteruption (B) photographs of Plot 485.

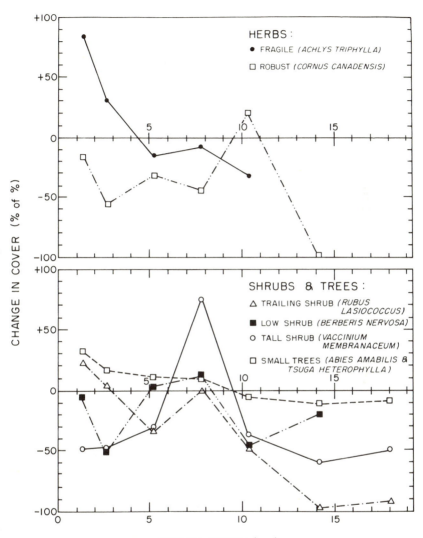

FIG. 7.10   Changes in percentage of cover from original preeruption measurements of seven vascular plant species on snow–free plots at Mount St. Helens.

photograph points, and (3) descriptions of the tephra deposits that can be updated as new eruptions and erosion or deposition occur. We will continue to perform visual estimates of percentage of cover by species

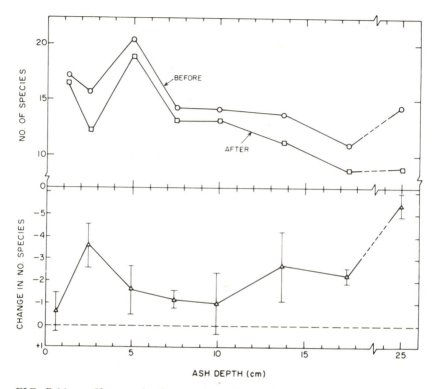

FIG. 7.11   Changes in the number of vascular plant species present on snow-free plots after the 1980 Mount St. Helens eruptions.

and make species lists for the entire plot. These measurements will allow long-term evolution of succession on the same land unit.

In spite of the difficulties with statistical analyses, several points emerge regarding the interaction of various disturbance types and intensities with the combination of vegetative and environmental conditions on our plots:

1. The presence of a snowpack on May 18, 1980, changed the effects of the blast and tephra deposition by (a) altering the physical characteristics of the tephra deposits, thus allowing easier penetration of sprouting plants, and (b) matting some shrub species to the ground, thus increasing the effectiveness of a given tephra depth in reducing shrub cover.

2. The physiognomy of plant communities affected the drop in vegetative richness and cover caused by a given ash depth. Herb-

dominated communities were generally more heavily impacted than shrub-dominated communities.

3. The stage of phenologic development (leaf-out) changed the impact of a given depth of tephra. Plant communities at lower elevations where buds had burst and leaves were more well developed were more heavily impacted than similar communities at higher elevations where most plants were still dormant.

Long-term remeasurement of these permanent plots can document a series of successions involving several plant communities. Such observations should address successional convergence, equilibria, resistance–resilience, and rates of recovery from impact (White 1979). Such disturbances as the eruptions of Mount St. Helens, which are relatively frequent in parts of the volcanic Cascades, play an important role in the geographic distribution and evolution of the flora in this region.

LITERATURE CITED

BORMAN, F. H., and G. E. LIKENS. 1979. *Pattern and Process in a Forested Ecosystem.* Springer-Verlag, New York, 253 pp.

BROCKWAY, D. G., M. A. HEMSTROM, and W. H. EMMINGHAM. 1982. *Preliminary Plant Association and Management Guide for the Pacific Silver Fir Zone of the Gifford Pinchot National Forest.* USDA Forest Service, Pacific Northwest Region, Gifford Pinchot National Forest, Vancouver, Washington, 122 pp.

HENRY, J. D., and J. M. A. SWAN. 1974. Reconstructing forest history from live and dead material: An approach to the study of forest succession in southwest New Hampshire. *Ecology* 55:772–783.

WHITE, P. S. 1979. Pattern, process, and natural disturbance in vegetation. *Bot. Rev.* 45(3):230–285.

TABLE 7.1

SUMMARY OF VEGETATION PLOTS RELOCATED IN SEPTEMBER, 1980,
NEAR MOUNT ST. HELENS

Number of plots	Disturbance type	Tephra depth (cm)	Relocation[a]
4	Mudflow	—	4 A
5	Blast	—	3 A, 2 E
1	Tephra	17.5	E
4	Tephra	15−17.5	4 E
6	Tephra	12.5−15	6 E
8	Tephra	10−12.5	4 A, 4 E
9	Tephra	7.5−10	3 A, 6 E
14	Tephra	5−7.5	3 A, 11 E
16	Tephra	2.5−5	3 A, 13 E
10	Tephra	2.5	2 A, 8 E

[a]A = approximately relocated; E = exactly relocated.

TABLE 7.2
ENVIRONMENTAL DATA, TEPHRA DEPTH AND COMPARISON OF DOMINANT PREERUPTION AND POSTERUPTION PLANT SPECIES PRESENCE AND PERCENTAGE OF COVER FOR EIGHT SAMPLE PLOTS NEAR MOUNT ST. HELENS[a]

	Plot 393		Plot 135		Plot 362		Plot 163		Plot 366		Plot 387		Plot 388		Plot 485	
	Before	After	Before	After	Before	After	Before	After	Before	After	Before	After	Before	After	Before	After
Elevation (m)	1170		1330		1000		1000		790		1300		1330		730	
Aspect (azimuth)	20		90		211		56		210		290		7		204	
Slope (%)	6		58		49		62		67		15		60		83	
Disturbance type[b]	M		B		B		B		B		T		T		T	
Tephra depth (cm)	0		20		5		13		25		13		4		1	
Total canopy cover (%)	70	0	60	0	90	0	80	0	90	0	60	55	45	35	90	80
Total shrub cover (%)	100	0	40	0	20	0	2	0	5	1	95	0	100	60	70	70
Total herb cover (%)	70	0	50	0	5	0	1	0	6	1	10	10	85	60	15	5
Trees																
Abies amabilis, mature	15	0	30	0	30	0	30	0	—[c]	—	30	30	45	35	—	—
A. amabilis, regeneration	10	0	35	0	8	0	1	0	—	—	15	12	15	15	—	—
A. procera	5	0	40	—	—	—	—	—	—	—	—	—	—	—	—	—
Pseudotsuga menziesii	30	0	—	—	15	0	20	0	50	0	—	—	—	—	60	50
Tsuga heterophylla, mature	20	0	—	—	45	0	20	0	40	0	10	10	—	—	20	20
T. heterophylla, regeneration	2	0	—	—	15	0	—	—	—	—	—	—	—	—	10	10
T. mertensiana	—	—	—	—	—	—	—	—	—	—	20	15	—	—	—	—

TABLE 7.2

ENVIRONMENTAL DATA, TEPHRA DEPTH AND COMPARISON OF DOMINANT PREERUPTION AND POSTERUPTION PLANT SPECIES PRESENCE AND PERCENTAGE OF COVER FOR EIGHT SAMPLE PLOTS NEAR MOUNT ST. HELENS[a] (continued)

	Plot 393		Plot 135		Plot 362		Plot 163		Plot 366		Plot 387		Plot 388		Plot 485	
	Before	After	Before	After	Before	After	Before	After	Before	After	Before	After	Before	After	Before	After
Tall Shrubs																
Acer circinatum	—	—	—	—	8	0	—	—	8	1	—	—	—	—	1	0
Gaultheria shallon	—	—	—	—	—	—	—	—	—	—	—	—	—	—	35	30
Menziesia ferruginea	5	0	6	0	—	—	0	1	—	—	40	5	8	3	—	—
Oplopanax horridum	—	—	—	—	—	—	—	—	—	—	—	—	40	40	—	—
Rhododendron albiflorum	—	—	—	—	—	—	—	—	—	—	25	3	—	—	—	—
Vaccinium spp.	60	0	35	0	9	—	3	3	4	0	18	4	8	2	—	—
Low Shrubs																
Berberis nervosa	—	—	—	—	—	—	—	—	2	1	—	—	—	—	20	20
Chimaphila umbellata	—	—	3	0	20	0	—	—	5	0	—	—	—	—	4	4

Trailing Shrubs

Species																	
Rubus ursinus	5	0	—	2	—	—	—	—	—	2	1	—	8	—	5	0	4	4
R. lasiococcus	10	0	2	0	—	1	—	—	0	1	—	—	8	2	5	0	—	—
Linnaea borealis	—	—	—	—	—	1	—	—	—	—	—	—	—	—	—	—	—	—
Achlys triphylla	—	—	3	0	1	—	—	—	—	—	—	4	4	2	—	—	4	—
Anemone deltoidea	—	—	8	0	0	1	—	—	—	—	—	—	—	—	—	—	4	0
Clintonia uniflora	25	0	8	0	0	—	—	—	—	—	—	—	—	—	—	—	—	—
Cornus canadensis	5	0	8	0	1	—	—	—	—	—	—	—	—	—	—	—	—	—
Erythronium montanum	—	—	—	—	—	—	—	—	—	—	0	10	0	—	—	—	—	—
Smilacina stellata	4	0	—	—	—	—	—	—	—	—	—	—	4	—	4	4	—	—
Tiarella trifoliata	20	0	15	0	1	0	—	—	0	1	—	—	4	—	4	4	0	1
Vancouveria hexandra	—	—	2	0	—	—	—	—	—	—	—	—	—	—	—	—	—	—
Viola sempervirens	—	—	—	—	—	—	—	—	—	—	—	—	—	—	—	—	—	—
Xerophyllum tenax	—	—	—	—	—	—	—	—	—	—	—	—	—	—	—	—	—	—

Ferns

Species																	
Atherium felix-femina	2	0	—	—	—	—	—	—	—	—	—	—	—	—	—	—	—	—
Gymnocarpium dryopteris	1	0	—	—	—	—	—	—	—	—	—	—	—	8	4	—	—	4
Polystichum munitum	2	0	3	0	1	0	—	—	—	—	—	—	—	0	0	2	10	10

[a] Original measurements were taken in the summers of 1978 and 1979, and remeasurements in August and September, 1980.

[b] M = mudflow, B = blast and tephra, T = tephra only.

[c] — indicates no data.

8

Revegetation in the Western Portion of the Mount St. Helens Blast Zone during 1980 and 1981

R. G. Stevens, J. K. Winjum,
R. R. Gilchrist, and D. A. Leslie

ABSTRACT

Revegetation on forest lands in the western portion of the Mount St. Helens blast zone following the eruption of May 18, 1980, was studied during the summers of 1980 and 1981. Tephra deposits ranged from 5 to 20 cm in depth across the study area. Study plots were in clearcut areas harvested <1–9 yr before May 1980. Both natural and artificial revegetation were investigated, the latter by planting bare-rooted Douglas fir and noble fir seedlings as well as sowing seeds of bentgrass, fescue, and rye grasses. Natural revegetation began to emerge through the tephra deposit in June, 1980, mainly by species sprouting shoots from surviving underground plant parts. Successful natural revegetation by seed on top of the tephra was rare because of the nitrogen deficiency in the tephra deposit. Vegetative cover, frequency, and height through 1980 and 1981 were much greater on the older clearcuts. By the fall of 1981, 31 species had been recorded within the study plots. Average vegetative cover ranged from 24 to 89% on 1-m² plots with some individual plots having as high as 100% cover. Douglas fir and noble fir seedlings planted with their roots in the underlying soil survived and grew better than those with their roots totally in the tephra deposit. Grass seeded on the tephra surface without NPK fertilizer germinated but died within a few days. When the fertilizer was used, rye grasses had by far the most successful establishment among the species tested.

INTRODUCTION

The eruption of Mount St. Helens on May 18, 1980, adversely affected the plant cover on forest lands north of the volcano in two significant ways. First, the lateral blast killed the above-ground parts of all terrestrial plants on 60,000 ha in a 160° arc with a radius as large as

25 km (Fig. 8.1). The blast zone had been a patchwork of mature conifer forests, young plantations, and areas recently harvested, that is, clearcuts. Underground plant parts such as roots, rhizomes, and dormant seed survived the blast in most outer areas of the zone. Second, the eruption of May 18 resulted in a heavy deposit of airfall tephra from a few centimeters to >40 cm deep on these same devastated lands. As a result the blast zone was left without any outward signs of life, causing many biologists to be concerned about future plant and animal existence. In addition, forest managers responsible for these lands were interested in their potential productivity for commercial conifer crops, one of the most dominant land uses prior to the eruption.

Francis (1976) described explosive volcanic eruptions as "displays of nature's most powerful forces at work." During the past century, botanists have studied and, in a few key instances, published descriptions of vegetative recovery on lands completely denuded by volcanic eruptions (Table 8.1). Although these botanical studies were conducted several years after the volcanic eruptions, most investigators were struck by the vigorous recovery of the native plants. Descriptions often included comments marveling at the resilience of nature (Nicholls 1963). In the ecological investigations, surviving stems and roots initiated recovery of local plant communities in the outer areas of most volcanically devastated areas where tephra deposits were shallow. Plants established by seed germinating on top of the tephra layer were rare during the first few growing seasons, mainly due to a deficiency in available nitrogen. Close to the volcanoes, recovery usually took decades and began with primary plant succession, gradually progressing toward the preeruptive plant communities.

The midspring timing of the Mount St. Helens eruption on May 18, 1980, offered a unique opportunity for observations on the recovery of plant growth within a few weeks of the devastation. This investigation will compare plant recovery in the outer northwest portion of the blast zone to that described for other volcanoes, as well as that occurring in the Pacific Northwest after clearcut logging. Revegetation of other areas of the blast zone, including the slopes of the volcano, are under study by several investigators (see, e.g., Chapters 5, 6, 7, and 9, this volume, and Winjum et al. 1982).

MATERIALS AND METHODS

Research sites were placed in the most recent clearcuts within a study area in that part of the Green River drainage that flows westward out of

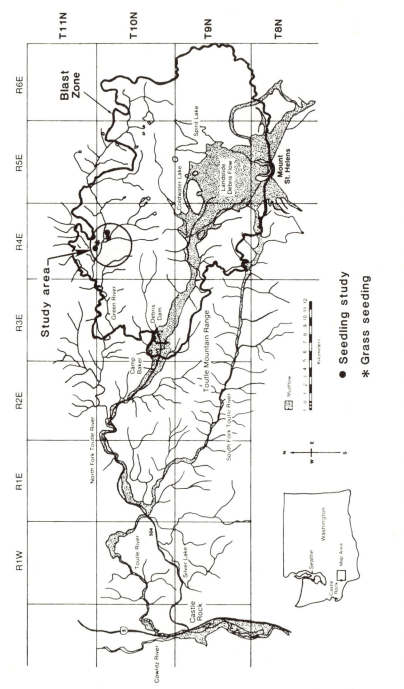

● Seedling study

＊ Grass seeding

FIG. 8.1   Location of Mount St. Helens blast zone in Washington and locations of plots within the blast zone.

the blast zone (Fig. 8.1 and Table 8.2). All sites are in T. 10N, R. 4E, Willamette meridian, and within 15–23 km of the volcano. The land is owned by Weyerhaeuser Company as part of its Southwest Washington Tree Farm. Conifer plantations ranging in age from 1 to 25 yr on ~10,000 ha were killed by the blast. In addition, 15,000 ha of forests with commercial-size trees were also destroyed. Topography of the area is mountainous, ranging from 500 to 1000 m in elevation. Major ridges are situated in a westerly direction away from the major north–south crest of the southern Cascade mountain range of Washington. Productivity of the land for conifer forests is moderately high. The area is included within the western hemlock (*Tsuga heterophylla*) zone (Franklin and Dyrness 1973), although natural forests and young growth plantations here are predominantly Douglas fir (*Pseudotsuga menziesii*). Annual precipitation is 200 cm/yr, occurring predominantly from November through April.

In the western portion of the zone, the tephra deposit that fell on May 18, 1980, is primarily the three-layered Unit A described by Waitt and Dzurisin (1981). The first layer (A1) is a gravel covered by a layer (A2) of sand. The depth of A1 plus A2 ranges from 5 to 20 cm. The top layer (A3) of fine silt is about 1.5 cm deep. This silt layer acted like a crust preventing infiltration of precipitation, such that intense erosion in the form of frequent and sharply cut rills occurred on the hillslopes with each rain event during the remainder of 1980.

Revegetation studies began in mid-June 1980 at five sites (Table 8.2) in clearcuts regenerated from 1971 to 1980 and denuded of vegetation by the volcanic blast in May 1980. All clearcuts had been burned prior to regeneration as part of standard forestry practices. To document natural revegetation, a system of 1-m² plots were established 10–50 m apart along five transect lines. Each line began at a randomly chosen point and ran perpendicular to the contours. Metal stakes at each plot enabled a 1-m² area to be identically relocated for periodic readings. The surface area inside the plot was not corrected for slope. The percentage of vegetative cover was estimated by taking a vertical projection of foliar material to the ground surface within the 1-m² plot and estimating the percentage of ground covered by vegetation. Each plant species (scientific names given in Table 8.3) within every plot was recorded periodically from July 1980 to October 1981. The height of the tallest plant in each plot was measured to the nearest centimeter.

Experiments to artificially establish plants by planting trees and sowing grass also began in mid-June 1980. For planting experiments, Douglas fir and noble fir (*Abies procera*) seedlings from a local seed source were still available in June 1980, although in a normal planting year seedlings would be planted or discarded by May 1. The Douglas fir were grown 2 yr in a nursery bed and outplanted as bare-root seedlings. The noble fir were 1-yr-old containerized seedlings grown in a greenhouse and planted with the containers and media removed. A few days before obtaining the noble fir seedlings, the peat and vermiculite plug around the roots had been removed for transplanting at the nursery. Therefore, the noble fir stock was a special type of bare-root seedling. The seedlings had been in conventional freezer storage ($-1°C$) since lifting in the nursery during February. Despite the less than optimum timing, it was thought that utilizing these seedlings could yield initial information vital for the planning of any large-scale plantings during the winter of 1980–81.

Two sites were chosen in the study area for seedling experiments (Fig. 8.1). For noble fir seedlings, three shovel planting treatments were tested: (1) roots were placed in undisturbed tephra, (2) roots were placed in underlying mineral soil after tephra was "scalped" away, and (3) roots were placed in a mix of tephra and soil. Mixing was accomplished by shoveling the tephra and underlying soil together into the planting hole. The limited number of Douglas fir seedlings permitted only treatments (1) and (2). At the two sites, each treatment consisted of two replicates of 50 seedlings spaced at 1-m intervals along a row. An additional 30 seedlings of each species were potted in media and placed in a greenhouse to assess potential performance of the stock.

Rainfall a few days before and during planting moistened the tephra deposit to a point where it was actually soft and slippery to walk on. Because the eruption occurred in midspring after abundant winter rains, the underlying soil was fully charged so that moisture was not a limiting factor. Periodic soil sampling indicated that without an abundant transpiring plant cover, available soil moisture in the underlying mineral soil remained high throughout the summer of 1980 and even through the 1981 growing season. All seedlings were periodically examined, and seedling survival, height, and vigor were recorded. Seedling vigor was determined by estimating the percentage of needles damaged or missing. The vigor scale ranged from 0 (no damage) to 4 (dead) with vigor 1 and 2 seedlings having <25% and 25–50% damage, respectively, and vigor 3 seedlings having >50% damage. Vigor 0, 1, and 2 seedlings are referred to as "high" vigor and have a high probability of survival.

Grass seed was sown on top of the tephra deposit at two locations (Fig. 8.1). The seed sown in four 10 × 10 m plots at each location was a mixture of the following species: red fescue (15%), highland colonial bentgrass (11%), perennial rye (24%), annual rye (24%), alta tall fescue

(23%), and inert material (3%) (scientific names given in Table 8.3). The sowing rate of the mix was 17 kg/ha. Half the plots received 8–8–8 NPK fertilizer applied by hand at the elemental rate for N of 18 kg/ha on the sowing date. Observations were made periodically after the June sowing and continued through the spring of 1981. Germination dates, grass condition, and estimated percentage of cover of subsample plots were recorded.

RESULTS

Natural vegetation, although not visible in the blast zone in early June, 1980, began to emerge in the northwest portion in mid-June (Table 8.4 and Fig. 8.2). For example, bracken fern, fireweed, and Canadian thistle shoots began sparingly then more conspicuously to penetrate the tephra. Emergence initially occurred in the shallower areas of the tephra near stumps or eroded rills. Other plants appeared in the polygonal cracks produced as the sun in late June began to dry out the tephra saturated by earlier rains.

From June through August 1980, the natural revegetation appeared to be from shoots produced by live plant parts such as roots and stems under the tephra layer (Table 8.4). The number of plant species emerging in June on the plot transects was <10, but by October 1980 the number reached 18 (Table 8.4). In a similar manner, the number of plots containing an individual species (i.e., species frequency) and the number of plots with any plant (i.e., plot frequency) increased from June to October 1980 (Table 8.4 and Fig. 8.2). Combining data from all 75 plots on the five transects, the average plot frequency increased from 33% in June to 74% in October (Fig. 8.2). Also, the percentage of cover increased from virtually nil in May to >50% in some locations by October (Fig. 8.3).

Clearcut age was strongly correlated to the degree of natural revegetation within the study area. By October 1980 the two transects that were placed in clearcuts <2 yr in age (burned May 1980 and October 1979) had vegetation growing in slightly more than half of the plots (Fig. 8.2). The older clearcuts (>2 yr of age) had vegetation occurring in 75–100% of the plots. This relationship was consistently shown in terms of percentage of cover (Fig. 8.3) and height of the tallest plants (Fig. 8.4) throughout 1980.

Most of the dicotyledonous plants flowered prolifically once the shoots matured. In October 1980, germinants of fireweed and Canadian thistle were observed on the surface of the tephra

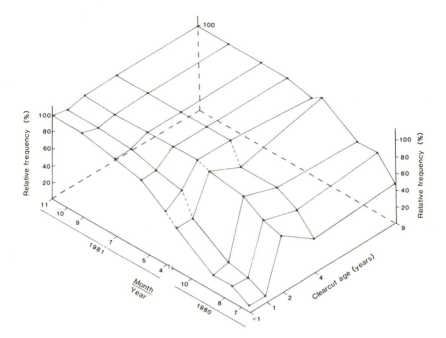

FIG. 8.2   The effect of clearcut age on the relative frequency of vegetation (percentage of plots with vegetation) during the first 2 yr following
the May 18, 1980, eruption.

(Table 8.4). Because of the distance to outside seed sources and the
quantity of germinants, it seems likely that plants within the blast zone
were the seed source. Few of the germinants survived the winter.

The USDA Soil Conservation Service aerially applied grass–
legume seed and NPK fertilizer to 8000 ha of the blast zone in the fall
of 1980 in an effort to minimize downstream damage from runoff and
erosion (Klock 1981, Stroh and Oyler 1981). Seeding on Weyerhaeuser
clearcuts was done between September 15 and October 20, 1980. One
clearcut in this study area (the <1-yr-old area) was not treated with
grass and fertilizer, but very little fall germination occurred on the
treated clearcuts. By spring the introduced grass species had germinated and were becoming a component of the total vegetative cover.

During 1981, the species richness on the plots continued to increase (Table 8.4), with 31 species recorded by October. The new
species appeared to be predominantly from remnants of plants within
the plot area that did not grow 1980 shoots after the volcanic blast but
remained viable under the tephra until the spring or summer of 1981.

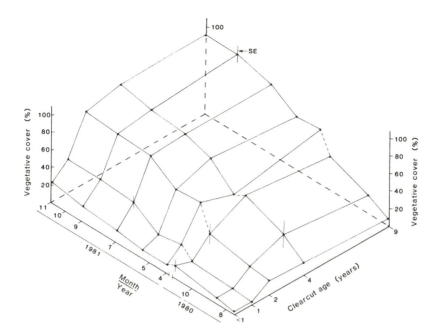

FIG. 8.3   The effect of clearcut age on the percentage of vegetative cover during the first 2 yr following the May 18, 1980, eruption. The bar at each age point indicates the maximum standard error of the mean recorded on that transect during the study.

The relative frequency of vegetation continued to increase during 1981, and even the youngest clearcuts approached 100% relative frequency (Fig. 8.2). The increase in vegetation was attributable to natural species on the <1-year-old clearcut that was not seeded. By September 1981 only 3 of the 18 plots on the 1-yr transect had grass occurring as the only vegetation within the plot. In all other plots on all transects, natural vegetation was present if grass occurred. This indicated (at least in the study area) that the seeded grass became established in environments that were also suitable for natural species to reestablish themselves.

The percentage of vegetative cover on the clearcuts continued to increase during 1981 (Fig. 8.3). The age of clearcut continued to be a significant factor. The <1-yr-old clearcut obtained only 24% average cover by October in contrast to the >90% cover on the 2-yr and older clearcuts. The 1-yr-old clearcut that was seeded and fertilized averaged 37% cover by September 1981.

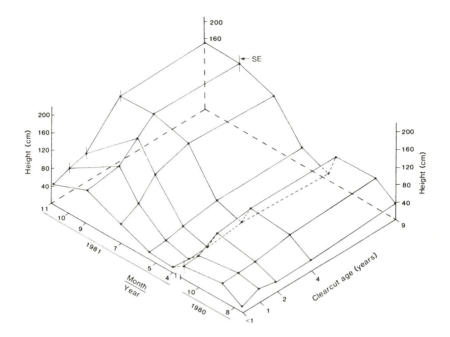

FIG. 8.4   The effect of clearcut age on the maximum height of vegetation (average of maximum vegetation height in transect plots) during
the first 2 yr following the May 18, 1980, eruption. The bar at each age
point indicates the maximum standard error of the mean recorded on
that transect during the study.

The 3 plots that had grass as the only vegetation had cover
ranging from 1 to 40%. Of the 18 plots in the transect, 4 plots did not
have grass established by September 1981. Because vegetative cover
was estimated as a vertical projection of foliar material to the ground,
only a total vegetative cover was determined. Field observations, however, indicated that on the 11 plots having grass and natural vegetation
occurring together, natural plants always represented a significant
percentage of the vegetative cover. On the 2-, 4-, and 9-yr-old clearcuts, introduced grass species occurred in almost all plots. Since native
vegetation had been better established on these transects in 1980,
however, the grass played a less significant role in total vegetative
cover. The grass mainly occurred as clumps under the larger native
species, and significant grass cover mainly occurred on the more gently
sloping 4- and 9-yr-old clearcuts. The predominant grass species noted
was annual rye grass.

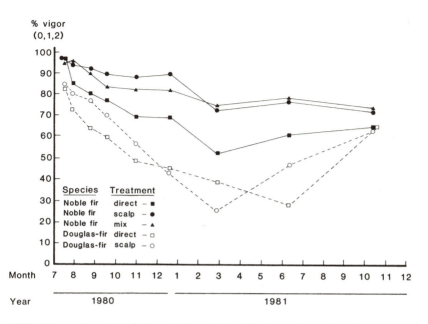

FIG. 8.5   Survival of vigor 0, 1, and 2 noble fir and Douglas fir seedlings planted in 10−20 cm of tephra by scalping, mixing with soil, or direct planting. Vigor 0−2 seedlings are those with< 50% damage or needle loss.

The height of the tallest vegetation in the plots also increased in 1981 (Fig. 8.4). The age of clearcut remained a factor, with older units having taller vegetation. The older clearcuts had more Canadian thistle and fireweed, which attained significantly more height than did senecio and other species occurring on the more recent clearcuts. Canadian thistle and fireweed taller than 3 m occurred in several areas. This luxuriant growth may have resulted partially from the fertilizer added with the grass seeding and from abundant soil moisture available under the tephra.

Artificial revegetation of these tephra-covered forest lands with conifer seedlings and grass seeding soon proved possible. With both conifers and grasses, however, success required overcoming the deficiency of available nitrogen of the Mount St. Helens tephra deposits (Moen and McLucas 1981).

The two sites selected for the seedling planting trials varied in tephra depth, with Site I (Fig. 8.5) being covered with generally >15 cm of tephra. The shallow tephra Site II (Fig. 8.6) had <10 cm of tephra. Noble fir survived better than Douglas fir (Figs. 8.5 and 8.6), which may have resulted from factors other than tephra, such as differences in seedling morphology, vigor, and responses to late planting date. A subsample of the seedlings potted in the greenhouse, however, had survival rates of 100%, indicating that the seedlings were viable at outplanting. Planting treatment did not affect the percentage of high vigor (vigor 0, 1, and 2) Douglas fir seedlings, although at the deeper tephra site (Fig. 8.5), scalping and mixing did improve the percentage of high vigor noble fir seedlings. This improvement was due to improved root contact with the underlying old forest soil where nitrogen was available. The early loss of vigor on the scalped noble fir treatment at the shallow tephra site (Fig. 8.6) resulted from weevil (*Steremnius* spp.) girdling at the ground line. The scalp environment was optimum for the weevils, and the residual population caused significant damage. Douglas fir seedlings with stem diameters >4 mm were not damaged by the weevils. Weevil girdling has not been a problem with establishment of operational conifer plantations because seedlings with large stem diameters >4 mm are used.

The sharp decline in seedling vigor from December 1980 to March 1981 at the shallow tephra site (Fig. 8.6) was due to browse

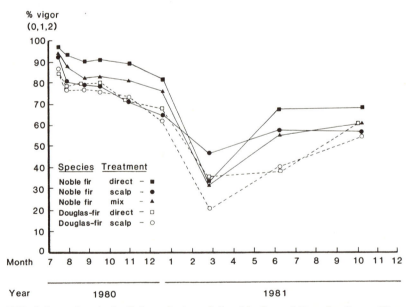

FIG. 8.6   Survival of vigor 0, 1, and 2 noble fir and Douglas fir seedlings planted in <10 cm of tephra by scalping, mixing with soil, or direct planting. Vigor 0−2 seedlings are those with <50% damage or needle loss.

damage caused by elk (*Cervus canadenses*). Although the seedlings were heavily browsed, they were able to flush in the spring and significantly improve their vigor. Areas of significant elk browse damage have been noted within the blast zone; here, extra large seedlings are planted at closer spacing, which minimizes losses from elk browsing.

Grass seed germinated on moist tephra after rain that occurred periodically throughout the summer of 1980. During periods of clear weather, however, the surface of the tephra quickly dried to form hard crust leaving ungerminated seeds lying on the surface. Without fertilization the germinants showed nutrient stress, and most of them died before their roots could penetrate the tephra layer to reach the nutrients in the forest soil below. In contrast, the fertilized grass plots developed a lush green cover by the fall of 1980 and were vigorous through 1981. Observations in the spring of 1981 showed that the grass species that performed best on fertilized plots were annual rye and perennial rye, with bentgrass doing moderately well. The fescues did not become established in significant numbers on any plots. At the assessment in the spring of 1981, the percentage of grass cover on unfertilized plots was estimated to be only 15%, whereas the fertilized plots had 75% cover.

DISCUSSION

The analyses of revegetation in 1980 gave important early evidence that the forest lands denuded by the lateral blast of Mount St. Helens and covered with <15 cm of tephra were almost immediately receptive to reestablishment by plants. Since the eruption was in midspring, the natural revegetation process began quickly because of the inherent readiness of plants to grow vigorously during that season. Although elementary in retrospect, this conclusion was reassuring to scientists and land managers totally unfamiliar with the restoration of volcanically devastated forest lands. Further support came from the literature of revegetation following volcanic eruptions in various locations of the world (Griggs 1918, 1919; Nicholls 1959; Bailey 1963).

Rees (1970) and others have noted that volcanic tephra is commonly without available nitrogen for plant growth. This unavailability is remedied by the gradual mineralization of the tephra layer (Francis 1976), which causes tephra to be mixed with soil (Griggs 1933), or by the introduction of nitrogen-fixing plants sometime during primary succession (Eggler 1959). The Mount St. Helens studies, however, indicate there are short-term solutions to revegetation of large areas covered with nitrogen-deficient tephra. Plants of the geophyte life form are capable of surviving beneath the tephra deposits. Through the late spring and summer of 1980, many plants near the study plots were dug to confirm that shoots were emerging from surviving root systems in the underlying soil. Thus, revegetation during the first few months of 1980 in the study area was from species under the tephra that had been present prior to the eruption.

A few germinants on top of the tephra were seen by the late fall of 1980. Successful germinants were commonly found in the disrupted areas of the tephra deposit that enlarged with time through disturbances from erosion, animals, and humans. The germinants represent another possible short-term solution to posteruptive revegetation. It is conceivable that viable seeds residing in the old litter layer could have germinated and grown through the tephra layer. Griggs (1918) proposed this possibility after his studies at Kodiak Island, Alaska, following the huge eruption of the Katmai Volcano in 1912. (Although the phenomenon was not observed in these studies, it was also not subjected to intensive investigation.) Seed also originated from the posteruptive 1980 flowering plants. In any event, the regenerated plant species were always the same as those previously present at the local site so that invasion by seed from long distances seemed unlikely. During 1981, natural plants continued to be predominantly from plants that emerged from beneath the tephra layer.

The natural regeneration in the shallow tephra areas of the Mount St. Helens blast zone represents a special type of secondary succession. Forest lands west of the Cascades in Oregon and Washington typically revegetate after fire or timber harvest by secondary succession patterns (Franklin and Dyrness 1973). In our study area, however, the early stage of secondary succession that commonly includes herbaceous invaders (Isaac 1940, Yerkes 1960, Dyrness 1973, Henderson 1978) was bypassed in favor of established geophytes.

Secondary succession on cleared land west of the Cascade range advances each year in abundance and toward more perennial plants (Morris 1958, Steen 1966, Bailey 1966). Thus it is not surprising that the age of clearcut before the eruption strongly influenced revegetation. The blast and tephra deposits temporarily delayed plant community development, screened out those species highly dependent on seed propagation, and let the geophytes continue after the eruption. Older clearcuts with their more advanced secondary succession had greater revegetation by the end of the first and second growing seasons than clearcuts <2 yrs old (Figs. 8.2–8.4).

The results from the outer blast zone having shallow tephra fit well with the data summarized by Rejmanek et al. (1982) for colonization of disturbed areas. The 31 species found after two growing seasons in this study are closer in magnitude to the 50 species estimated for postlogging succession in Oregon than the <10 species estimated for cinder cones of Paricutin and Puu Paui volcanoes. The prevalence of native plants reestablishing themselves from existing plant parts caused succession to be more advanced than on cinder cones. Colonization and succession in deeper tephra areas of Mount St. Helens may follow colonization curves that more closely resemble those for cinder cones where introduced species play a more important role.

No reports exist in the literature describing the artificial revegetation of volcanically devastated areas during the first year. The successful planting of Douglas fir and noble fir was significantly important for the future management of western blast zone lands. These two conifers are highly valuable commercial tree species in western Washington and Oregon, and regeneration techniques for them are well known. As a result, 16,000 ha of western blast zone lands have been scheduled for planting with these tree species from 1981 through 1985.

The introduced grasses proved to be very useful test plants for studying revegetation of the new tephra deposits. By August, 1980, results from the grass plots showed the need to consider the nitrogen deficiency of tephra in the artificial revegetation treatments in the blast zone.

In less than a day in May 1980, the lateral blast of the Mount St. Helens volcano changed a vigorous forest ecosystem into a seemingly gray wasteland. However, these early studies of revegetation in the western blast zone were successful in proving otherwise. Results from side-by-side studies of natural and artificial revegetation were complimentary and demonstrated the rapid recovery possible using the combined effort of nature and humans.

LITERATURE CITED

BAILEY, A. W. 1966. Forest associations and secondary plant succession in the southern Oregon Coast Range. Ph.D dissertation, Oregon State Univ., Corvallis.

BAILEY, W. H. 1963. Revegetation in the 1914–1915 devastated area of Lassen Volcanic National Park. Ph.D dissertation, Oregon State Univ., Corvallis.

DYRNESS, C. T. 1973. Early stages of plant succession following logging and burning in the western Cascades of Oregon. *Ecology* 54:57–69.

EGGLER, W. A. 1959. Manner of invasion of volcanic deposits by plants, with further evidence from Paricutin and Joruillo. *Ecology Monog.* 29(3):267–284.

FRANCIS, P. 1976. *Volcanos*. Penguin Books, New York. 368 pp.

FRANKLIN, J. F., and C. T. DYRNESS. 1973. *Natural Vegetation of Oregon and Washington*. PNW Forest and Range Exp. Station, USDA Forest Service, Gen. Tech. Report PNW-8. 417 pp.

GRIGGS, R. F. 1918. The recovery of vegetation at Kodiak. *Ohio J. Sci.* 19:1–57.

———. 1919. The beginnings of revegetation in Katmai valley. *Ohio J. Sci.* 19:318–342.

———. 1933. The colonization of the Katmai ash, a new and inorganic soil. *Amer. J. Bot.* 20:92–113.

HENDERSON, J. A. 1978. Plant succession on the *Alnus rubra/Rubus spectabilis* habitat type in western Oregon. *NW Sci.* 52(3):156–167.

HITCHCOCK, L. C., and A. CRONQUIST. 1973. *Flora of the Pacific Northwest.* Univ. Washington Press, Seattle. 730 pp.

HOWARD, R. A. 1962. Volcanism and vegetation in the Lesser Antilles. *J. Arnold Arbor* 18:279–315.

ISAAC, L. A. 1940. Vegetative succession following logging in the Douglas-fir region with special reference to fire. *For. J.* 38:716–721.

KLOCK, G. O. 1981. *A First-year Evaluation of the Erosion Control Seeding and Fertilization near Mount St. Helens.* A final report prepared for USDA Soil Conserv. Service, Spokane, Washington. 17 pp.

MOEN, W. S., and G. B. McLUCAS. 1981. *Mount St. Helens Ash: Properties and Possible Uses.* Washington State Dept. Nat. Res. Report of Investigations 24. 60 pp.

MORRIS, W. G. 1958. *Influence of Slash Burning on Regeneration, Other Plant Cover and Fire Hazard in the Douglas-Fir Region (A Progress Report).* PNW Forest and Range Exp. Station, USDA Forest Service, Res. Paper 29. 49 pp.

NICHOLLS, J. L. 1959. The volcanic eruptions of Mt. Tarawera and Lake Rotomahana and effects on surrounding forest. *N.Z. J. For.* 8:133–142.

———. 1963. Vulcanicity and indigenous vegetation in the Rotorua District. *N.Z. Ecol. Soc. Proc.* 10:58–65.

REES, J. D. 1970. Paricutin revisited: A review of man's attempts to adapt to ecological changes resulting from volcanic catastrophe. *Geoforum (Oxford)* 4:7–25.

REJMANEK, M., R. HAAGEROVA, and J. HAAGER. 1982. Progress of plant succession on the Paricutin volcano: 25 years after activity ceased. *Amer. Mid. Nat.* 108(1):194–199.

STEEN, H. K. 1966. Vegetation following slash fires in one western Oregon locality. *NW Sci.* 40(3):113–120.

STROH, J. R., and J. A. OYLER. 1981. SCS seeding evaluation on Mount St. Helens: Assessment of grass–legume seedings on Mount St. Helens blast area and the lower Toutle River mudflow, May, 1981. USDA Soil Conserv. Service, Spokane, Washington. 15 pp.

WAITT, R. B., Jr., and D. DZURISIN. 1981. Proximal air-fall deposits from the May 18 eruption: Stratigraphy and field sedimentology. In *The 1980 Eruptions of Mt. St. Helens, Washington,* Lipman, P. W., and D. R. Mullineaux (eds.), Geol. Surv. Prof. Paper 1250, pp. 601–616.

WINJUM, J. K., H. W. ANDERSON, and G. A. COOPER. 1982. Research associated with Mt. St. Helens and the volcanic eruptions of 1980. St. Helens Forest Land Research Coop., with Weyerhaeuser Co., Centralia, Washington. 129 pp.

YERKES, V. P. 1960. *Occurrence of Shrubs and Herbaceous Vegetation after Clearcutting Old-Growth Douglas-Fir in the Oregon Cascades.* PNW Forest and Range Exp. Station, USDA Forest Service Res. Paper 34. 12 pp.

TABLE 8.1
SELECTED LITERATURE ON REVEGETATION OF VOLCANICAL
DEVASTATED LANDS

Botanist[a]	Publishing year[a]	Volcanoes	Eruption years
Bailey, W. H.	1963	Lassen Peak, California	1914 and 1915
Eggler, W. A.	1959 ⎫	Paricutin and	1943–1952 and
Rejmanek, M. et al.	1982 ⎭	Jorullo, Mexico	1759–1775
Griggs, R. F.	1918, 1919, and 1933	Katmai, Alaska	1912
Howard, R. A.	1962	Mount Pelee, Soufriere, and others in the Lesser Antilles Is.	1902–1962
Nicholls, J. L.	1959 and 1963	Mount Tarawera, New Zealand	1886

[a] See literature cited for complete references.

TABLE 8.2
PHYSICAL CHARACTERISTICS OF CLEARCUT AREAS CONTAINING
THE FIVE NATURAL REVEGETATION STUDY SITES

Clearcut age (yr)	Number of plots	Slope (%)	Aspect	Tephra depth (cm)[a]	
				Fine layer	Total
9	15	5–15	N	4.4 ± 0.2	11.2 ± 0.5
4	8	0–15	N	3.6 ± 0.3	8.4 ± 0.6
2	19	25–60	NE, E, SE	2.9 ± 0.3	7.6 ± 0.6
1	18	35–50	NW	3.3 ± 0.1	7.9 ± 0.3
<1[b]	15	40–50	S	5.1 ± 0.3	9.6 ± 0.5
Total	75				

[a] Values are mean ± S.E.
[b] Burned May 1, 1980.

TABLE 8.3
Common and Scientific Names of Plants Mentioned
in This Chapter

Common name	Scientific name[a]

Species of natural vegetation observed on plot transects from June 1980 to October 1981

Black cottonwood	*Populus trichocarpa* T. & G.
Bleedingheart	*Dicentra* spp.
Bracken fern	*Pteridium aquilinum* (L.) Kuhn.
Bull thistle	*Cirsium vulgare* (Savi) Tenore
Canadian thistle	*Cirsium arvense* (L.) Scop.
Clover	*Trifolium* spp.
False dandelion	*Hypochaeris radicata* L.
Fireweed	*Epilobium angustifolium* L.
Grasses	Gramineae Family
Horsetail	*Equisetum* spp.
Lady fern	*Athyrium* spp.
Mosses	Bryophyta Division
Nettles	*Urtica* spp.
Oregon grape	*Berberis nervosa* Pursh
Oxalis	*Oxalis oregana* Nutt.
Pearly everlasting	*Anaphalis margaritacea* (L.) B. & H.
Red elderberry	*Sambucus racemosa* var. *arborescens* (T. & G.) Gray
Salal	*Gaultheria shallon* Pursh
Salmonberry	*Rubus spectabilis* Pursh
Senecio	*Senecio* spp.
Skunk cabbage	*Lysichitum* Schott
Solomon's seal, False	*Smilacina stellata* (L.) Desf.
Sweet coltsfoot	*Petasites frigidus* var. *palmatus* (Ait.) Crong.
Sword fern	*Polystichum munitum* (Kaulf.) Presl
Thimbleberry	*Rubus parviflorus* Nutt.
Trailing blackberry	*Rubus ursinus* Cham. & Schlecht.
Trillium	*Trillium ovatum* Pursh
Twisted stalk	*Streptopus* spp. Michx.
Vanilla leaf	*Achlys triphylla* (Smith) DC.
Vine maple	*Acer circinatum* Pursh
Willow	*Salix* spp.

Grass species used in seeding tests

Alta tall fescue	*Festuca arundinacea* Schreb
Annual rye	*Lolium multiflorum* Lam.
Highland colonial bentgrass	*Agrostis tenuis* Sibth
Perennial rye	*Lolium perenne* L.
Red fescue	*Festuca rubra* L.

[a]From Hitchcock and Cronquist 1973.

TABLE 8.4

THE RELATIVE FREQUENCY OF OCCURRENCE (% PLOTS HAVING THE
SPECIES) BY PLANT SPECIES FROM JUNE 1980 TO OCTOBER 1981

Species[a]	1980				1981		
	June 18	July 15	Aug. 19	Oct. 29	May 13	Sept. 1	Oct. 14
Fireweed	8	15	10	15	23	21	16
Canadian thistle	16	21	25	27	46	55	51
Bracken fern	7	8	8	15	12	21	18
False dandelion	5	6	7	7	11	10	8
Pearly everlasting	0	7	6	10	11	8	11
Trailing blackberry	3	6	10	12	12	14	19
Grasses	4	3	6	6	72	72	77
Horsetail	4	4	4	4	4	4	4
Senecio	1	4	6	10	18	14	2
Oregon grape	0	0	0	10	9	14	16
Thimbleberry	0	0	6	6	5	3	8
Vine maple	0	4	1	1	1	1	1
Clover	0	0	0	0	20	30	37
Oxalis	0	0	0	1	7	1	3
Salmonberry	0	3	0	1	1	3	1
Sweet coltsfoot	0	0	0	0	3	0	0
Sword fern	3	1	0	0	0	3	0
Red elderberry	0	0	0	0	1	1	0
Black cottonwood	0	0	0	1	0	0	3
Willow	0	0	0	0	3	3	3
Trillium	0	0	0	0	3	0	0
Salal	0	0	0	0	1	1	1
Mosses	0	0	0	0	8	0	0
Vanilla leaf	0	0	0	0	0	0	1
Bull thistle	0	0	0	0	0	0	1
Bleeding heart	0	0	0	0	1	0	0
Solomon's seal	0	0	0	0	3	1	0
Nettles	0	0	1	1	0	1	3
Lady fern	0	1	1	0	1	0	0
Twisted stalk	0	0	0	0	1	1	1
Skunk cabbage	0	0	0	0	3	0	0
Dicot germinants	0	0	0	29	9	1	18

[a] Scientific names given in Table 8.3.

9

Response of Vegetation within the Blast Zone, Mount St. Helens

Stephen D. Veirs, Jr.

ABSTRACT

The explosive eruption of Mount St. Helens on May 18, 1980, devastated surrounding above-ground vegetation through direct blast effects and deposition of ash and heated materials derived from the former mountain top. Following the eruption, a zone of blast-destroyed forest surrounded by a zone of standing dead trees was created to the west, north, and east of the mountain. Temperatures of the blast material that killed the above-ground vegetation in the standing dead zone can be inferred from aspects of the survival of the bigleaf maple (*Acer macrophyllum*) and other plants. Ferns, horsetails, biennials in the rosette stage, grasses, and other plants with underground stem tips survived extensively in the blast zone. Of the woody species, only those capable of sprouting from basal, buried portions of stems survived. Conifers were eliminated from large areas of former dominance.

INTRODUCTION

The explosive eruption of Mount St. Helens on May 18, 1980, appeared to reset ecological clocks when it devastated biological communities in a 600-km^2 area of southwestern Washington (Fig. 9.1). Eyewitness descriptions of the eruptive events are presented in detail by Rosenbaum and Waitt (1981). Vegetation north, northeast, and northwest of Mount St. Helens was destroyed by landslide, debris avalanche, and direct blast effects (Kieffer 1981, Moore and Sisson 1981, Wiatt 1981), by envelopment by hot gases (Winner and Casadevall 1981), and by deposition of hot blast materials (Banks and Hoblitt 1981, Hoblitt et al. 1981). The debris avalanche (Voight et al. 1981), coming to rest in the North Fork Toutle River, obliterated ~7,000 ha of forest and stream communities. About 35,000 ha of forest and recently logged forest vegetation was destroyed by direct

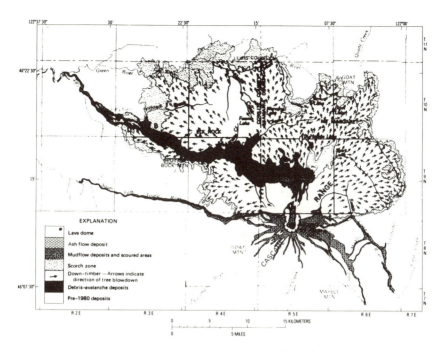

FIG. 9.1   Map of devastated area and study site locations: (D) Deer Creek, (B) Bear Creek, (M) Maratta Creek, (J) Jackson Creek, and (C) Castle Creek ridge study sites. (Modified from Fig. 257 in Wiatt 1981.)

blast effects (Figs. 9.2 and 9.3), and ~11,600 ha of forest and plant communities reestablished following logging were killed but not blown down (Fig. 9.4) (areas from Means et al. 1982). This scorch zone of standing dead vegetation formed a narrow ring from a few meters to 4 km in width around the blast zone to the north, northeast, and northwest of the summit.

This chapter discusses the results of a September 1980 reconnaissance of the vegetation found in the blast, avalanche deposit, and scorch zones in the North Fork Toutle River basin, northwest of the summit of Mount St. Helens (Fig. 9.1). Other studies of vegetation to the northeast of the volcano carried out in September 1980 are reported by Means et al. (1982). The areas of ash flow deposits north of the crater and mudflow deposits beyond the limits of the devastated area were not examined in the present study. The terms used here for the various deposits and blast effect zones are generally those of Wiatt (1981).

STUDY SITE AND METHODS

A helicopter provided access to a series of stations along the North Fork Toutle River valley upstream from Camp Baker to a ridge between Castle and Studebaker creeks. This access permitted a reconnaissance examination of living vegetation found in the areas of the scorch zone, the blast zone, the debris avalanche deposit, and its marginal levees. Five hillslope locations above the limits of the debris avalanche deposit were selected for detailed examination. Study site locations are shown in Fig. 9.1; physical descriptions of the sites are given in Table 9.1. The sites were located in recently cutover forest land and forest in the scorch (standing dead) zone (Deer and Bear creeks); in old-growth, young second growth, and recently clearcut land in down timber in the blast zone (Jackson Creek); and in second growth forest vegetation in the blast zone where trees were destroyed (Maratta Creek). The Castle Creek ridge site was in the down-timber zone, below the probable 1200 m lower limit of the snow pack as of May

FIG. 9.2   Blast-destroyed forests ~18 km northwest of the summit of Mount St. Helens, near Elk Rock. The logging road serves as a scale.

FIG. 9.3   Near view of blast-destroyed forest in the Jackson Creek study site,
~15 km northwest of the summit. Direct blast deposits and tephra mantle the
ground and forest debris.

18, 1980, reported by Antos and Zobel (1982). The natural upland
forest vegetation of the sites was dominated by *Pseudotsuga menziesii*,
Tsuga heterophylla, and *Thuja plicata* (the *Tsuga heterophylla* zone of
Franklin and Dyrness 1973). The Castle Creek ridge site may have
been in the *Abies amabilis* zone of Franklin and Dyrness (1973) as
indicated by the presence of *Alnus sinuata* in the second growth forest.

   At each of the five sites, living plants were identified or collected
for later identification. Their apparent origin (seed or sprout) and
relative location with respect to preeruption vegetation and topography and to posteruption deposits, depth of deposit, and soil moisture
conditions were noted. (Nomenclature follows Hitchcock and Cronquist 1973.) Walking traverses of the debris avalanche deposit and its
marginal levees were made adjacent to each site, except at Castle Creek
ridge, and at numerous other points on the debris avalanche deposit
west – northwest of the ash flow deposits. The Bear Creek site included
a traverse of the scorch zone along the Weyerhaeuser 3130 Road from
the green vegetation on the outer margin to the edge of the down-
timber zone.

FIG. 9.4   Standing dead conifers in the scorch zone, Bear Creek site, ~21 km northwest of the summit.

RESULTS AND DISCUSSION

A list of the living plants found at each site is given in Table 9.2. For each taxon the pre- and posteruption habitat and life form are described.

DEBRIS AVALANCHE DEPOSIT

With one exception, no living plants were found on the debris avalanche deposit in spite of the presence of large amounts of organic debris from preeruption vegetation. A few young seedlings of *Cirsium* sp. in the cotyledon stage were identified by the still-present pappus and were found in a tiny cavity in a jumbled mound of avalanche debris. They were rooted in tephra deposited after the active movement of the debris avalanche, and germination probably occurred following light rains that fell during the field work.

MARGINAL LEVEES

During the debris avalanche movement, marginal levees were formed. In addition to the avalanche material, these levees contained light brown sandy soil and plant fragments locally incorporated by the passing debris flow; they were mantled with a thin layer (1–2 cm) of tephra. The plant fragments in the levees did not appear to have been entrained in the main debris flow for any appreciable period. Sprouting of remnants of *Salix*, *Rubus spectabilis*, *R. ursinus*, *Sambucus*, and *Vaccinium parviflorum* had given rise to well-established, rooted plants. In addition to these sprouting species, scattered seedlings of *Salix* and *Lupinus* plus seedlings of preeruption plants of *Montia*, *Rumex*, and several grasses were found in the marginal levees. Presumably all of these plants were derived from local preeruption plant fragments or seed, with the possible exception of the *Salix*, which is easily windblown.

BLAST ZONE

At the Jackson Creek, Marrata Creek, and Castle Creek ridge sites, living plants covered less than an estimated 5% of the blast deposits. Each site had been completely or partially clearcut prior to the eruption. All taxa present on these former clearcut sites are known to sprout from rhizomes or dormant buds, or have a basal rosette. All are biennial or perennial. The most commonly observed plants on the clearcut sites included *Epilobium angustifolium* in relatively extensive stands, *Anaphalis margaritacea* (Fig. 9.5), *Rubus ursinus*, and *Cirsium*. Taxa associated with blown-down, old-growth, and advanced second growth forests on the Jackson Creek site were typical of the forest understory and included ferns, *Achlystriphylla*, *Oxalis oregona* (Fig. 9.6), and *Vaccinium parviflorum*. These survived in blast deposits 5–15 cm in depth on the Jackson Creek site (Fig. 9.7).

Preeruption seeps in old-growth and clearcut areas supported vigorous stands of *Salix*, *Equisetum* (Fig. 9.8), *Lysichitum americanum*, and *Petasites frigidus* having local cover as high as 25%. Other local microtopographic features, including preeruption root mounds (Fig. 9.9), root wads, and similar features with steep-sided surfaces that shed hot ash and blast debris, had greater plant survival than did surfaces where deposits were deeper. Trees surviving on the study sites were limited to posteruption basal sprouts of *Salix*, *Alnus sinuata*, *Acer macrophyllum*, and *A. circinatum*. No living conifers were seen on or around any of the study sites, but a few small trees were seen at higher elevations where snow pack protected them (Means et al. 1982).

FIG. 9.5   *Salix, Rubus parviflorus, Anaphalis margaritacea,* and *Epilobium an-gustifolium* vegetatively sprouting on the slope of Castle Creek ridge, above the debris avalanche deposit, ~8 km northwest of the summit.

Although relatively close (7 km) to the volcano on Castle Creek ridge, litter, soil moisture, soil cover, or microrelief afforded sufficient protection to permit widespread survival of individual plants with a rhizomatous or basal sprouting habit (Fig. 9.5). No living aerial portions of any plants that predated the blast were observed. Virtually all above-ground plant parts had been killed during the explosive phase of the eruption by the direct effects of breakage, abrasion, and heating.

SCORCH ZONE

Several names have been applied to the zone of standing dead forest surrounding the blast or down-timber zones: singed zone (Kieffer 1981), seared zone (Moore and Sisson 1981), and scorch zone (Means et al. 1982). In this scorch zone, in old growth at the Bear Creek site, all the overstory conifers (*Pseudotsuga menziesii, Tsuga heterophylla,* and *Thuja plicata*) were dead. In the same area living understory plants, especially those protected by high groundwater levels, micro-topography, or favorable growth habit, became increasingly numerous toward the outer margin of the zone (here ~1.5 km wide). A remark-

FIG. 9.6   Old-growth forest understory plants developing from stem tips that survived the eruption of May 18, 1980. Shown here are *Polystichum munitum*, *Athyrium felix-femina*, and *Oxalis oregana*.

able aspect of this zone was the presence of living *Acer macrophyllum* individuals that were at least 10–20 m tall and were sparsely branched and foliated along much of the length of their stems (Fig. 9.10). On closer inspection, it became apparent that the leaves were born on epicormic branches that developed after the May 18 eruption. The transient heating episode that caused the death of the preeruption foliage, branches, and stem tips of the *A. macrophyllum* did not kill dormant buds on the main stem or the cambium on the main stem, even in this relatively thin-barked species. The same exposure killed all of the conifers in the stand, even those with thicker bark. Basal (coppice) sprouting is common in *Acer* and hardwoods of many genera and in *Sequoia sempervirens*, but is rare in other conifers. Although heating may have been insufficient to kill the cambium of the standing dead conifers, no epicormic sprouting was seen and the trees appeared dead. This may have been due in part to the strong apical dominance and suppression of dormant bud development common in conifers (Kramer and Kozlowski 1979).

    On the basis of exposure of fir (*Abies amabilis*) leaves to heat, Winner and Casadevall (1981) placed the killing temperatures between

FIG. 9.7   A typical soil core from the Jackson Creek site. About 1 cm of tephra overlays 10 cm of blast debris and the original soil, ~15 km from the summit. (Scale in inches.)

50 and 250°C in this zone. The eyewitness accounts reported by Rosenbaum and Waitt (1981) and evidence presented by Banks and Hoblitt (1981) suggest that exposure to the elevated air temperature was brief, 1−2 min. but not >10 min, probably because of the onset of cool air flow from beyond the scorch zone toward the base of the ascending eruptive column. The vertical tree stems also must have been spared much of the adverse heating effects of the heated rock fragments and ash that settled on leaves and horizontal branches following the blast. In any case, it is apparent that the exposure to high temperatures within the scorch zone was not great enough to kill the cambium and dormant aerial buds in the relatively thin-barked main stems of *Acer macrophyllum*, but it was great enough to kill the same dormant buds or the cambium on smaller branches.

The eyewitness accounts and the relatively large number of living plants clearly show that superheated water vapor was not the killing agent in the blast and scorch zones. Live steam and the large associated change-of-state heat loss was expected to produce greater mortality in more exposed microhabitats. Instead, mortality (at least of understory plants) was greatest where blast deposits were deepest. This is con-

FIG. 9.8   Recovery of vegetation surviving in a seep on the Jackson Creek site. Plants include *Salix, Equisetum, Polystichum munitum, Mimulus moschatus,* and *Stachys mexicana.*

sistent with the interpretation that tree mortality in the scorch zone and understory plant mortality in the blast and scorch zones (of plants that escaped the direct blast effects and transient atmospheric heating) was a result of direct heat transfer from the hot rock fragments and ash deposited on them. Plant survival patterns in the scorch zone support the conclusion that all of the plants seen outside the marginal levees were survivors from preexisting vegetation. Survival was largely limited to basal sprouters, rhizomatous perennials, rosette-forming biennials, and other hemicryptophytes. Soil, duff, litter, and aerial plant parts must have provided some insulation from the effect of heated material settling on and around the plants. Survival in the scorch zone was enhanced, as elsewhere, by favorable microtopography and soil moisture conditions. The pattern of survival— old-growth understory species in the old-growth preeruption habitat and weedy species on preeruption road cut and fill surfaces and on recent clearcuts—strongly suggests that the living plants present in this zone were vegetatively derived from plants that survived the effects of the eruption.

FIG. 9.9   A root mound—an example of a microtopographic feature that favored survival. Plants are grasses, *Cirsium*, and *Rubus spectabilis*. Jackson Creek site.

CONCLUSIONS

The new materials created by the May 18, 1980, eruption of Mount St. Helens and subsequent eruptive activity, such as the pyroclastic flows, represent new habitats for ecological succession if there are no further volcanic disturbances. The heated, pulverized, older material of the debris avalanche deposit also created surfaces on which new successional sequences can occur. No higher plants survived there. In the areas examined in the blast and scorch zones northwest of the summit, relatively large numbers of individual plants representing a broad spectrum of species survived the devastating effects of the eruption. The near total loss of coniferous trees is the most notable change in the natural vegetation in the devastated area. The effects of the eruption of Mount St. Helens provide a major opportunity for investigations of the successional dynamics of plants.

FIG. 9.10   New foliage on epicormic branches from the stem of *Acer macrophyllum*. These trees survived among the standing dead conifers in the scorch zone, ~21 km from the summit.

ACKNOWLEDGMENTS

I gratefully acknowledge the logistical assistance of the U.S. Geological Survey staff, Vancouver Field Office (D. Peterson, Scientist in Charge), given during the course of field studies directed by R. J. Janda. I appreciate the assistance of J. M. Lenihan, Redwood National Park, Arcata, California, in plant identifications and confirmations.

LITERATURE CITED

ANTOS, J. A., and D. B. ZOBEL. 1982. Snowpack modifications of volcanic tephra effects on forest understory plants near Mount St. Helens. *Ecology* 63:1969–1972.

BANKS, N. G., and R. P. HOBLITT. 1981. Summary of temperature studies of 1980 deposits. In *The 1980 Eruptions of Mount St. Helens, Washington*, Lipman, P. W., and D. R. Mullineaux (eds.), U.S. Geol. Surv. Prof. Paper 1250, pp. 295–314.

FRANKLIN, F., and C. T. DYRNESS. 1973. *Natural Vegetation of Oregon and Washington.* USDA Forest Service Gen. Tech. Report PNW-8. Pacific Northwest Forest and Range Exp. Station.

HITCHCOCK, C. L., and A. CRONQUIST. 1973. *Flora of the Pacific Northwest.* Univ. of Washington Press, Seattle. 730 pp.

HOBLITT, R. P., C. D. MILLER, and J. W. VALLANCE. 1981. Origin and stratigraphy of the deposit produced by the May 18 directed blast. In *The 1980 Eruptions of Mount St. Helens, Washington*, Lipman, P. W., and D. R. Mullineaux (eds.), U.S. Geol. Surv. Prof. Paper 1250, pp. 401–420.

KIEFFER, S. W. 1981. Fluid dynamics of the May 18 blast at Mount St. Helens. In *The 1980 Eruptions of Mount St. Helens, Washington*, Lipman, P. W., and D. R. Mullineaux (eds.), U.S. Geol. Surv. Prof. Paper 1250, pp. 389–400.

KRAMER, P. J., and T. T. KOZLOWSKI. 1979. *Physiology of Woody Plants.* Academic Press, New York. 811 pp.

MEANS, E., W. A. MCKEE, W. H. MOIR, and F. FRANKLIN. 1982. Natural revegetation of the northeastern portion of the devastated area. In *Mount St. Helens: One Year Later*, S. A. C. Keller (ed.), Proc. of Symposium, May 17–18, 1981, Eastern Washington Univ., Cheney, 93–103.

MOORE, J. G., and T. W. SISSON. 1981. Deposits and effects on the May 18 pyroclastic surge. In *The 1980 Eruptions of Mount St. Helens, Washington*, Lipman, P. W., and D. R. Mullineaux (eds.), U.S. Geol. Surv. Prof. Paper 1250, pp. 421–438.

ROSENBAUM, J. G., and R. B. WIATT, JR. 1981. Summary of eyewitness accounts of the May 18 eruption. In *The 1980 Eruptions of Mount St. Helens, Washington*, Lipman, P. W., and D. R. Mullineaux (eds.), U.S. Geol. Surv. Prof. Paper 1250, pp. 53–68.

VOIGHT, B., H. GLICKEN, R. J. JANDA, and P. M. DOUGLAS. 1981. Catastrophic rockslide avalanche of May 18. In *The 1980 Eruptions of Mount St. Helens, Washington*, Lipman, P. W., and D. R. Mullineaux (eds.), U.S. Geol. Surv. Prof. Paper 1250, pp. 347–378.

WIATT, R. B., JR. 1981. Devastating pyroclastic density flow and attendant airfall of May 18: Stratigraphy and sedimentology of deposits. In *The 1980 Eruptions of Mount St. Helens, Washington*, Lipman, P. W., and D. R. Mullineaux (eds.), U.S. Geol. Surv. Prof. Paper 1250, pp. 439–460.

WINNER, W. E., and T. J. CASADEVALL. 1981. Fir leaves as thermometers during the May 18 eruption. In *The 1980 Eruptions of Mount St. Helens, Washington*, Lipman, P. W., and D. R. Mullineaux (eds.), U.S. Geol. Surv. Prof. Paper 1250, pp. 315–322.

TABLE 9.1

NAMES, LOCATIONS, AND PHYSICAL CHARACTERISTICS OF STUDY SITES
IN THE NORTH FORK TOUTLE DRAINAGE, WASHINGTON

Site name	Location	Elevation (m)	Distance (km) from summit of Mount St. Helens	Azimuth from summit to site	Site aspect
Bear Creek	Sec. 34 and 35, T.10N, R.3E. Elk Rock, Washington 15' Quad	450–670	21	306° NW	SSW-facing slope nearly parallel to blast. Scorch zone/down-timber zone.
Jackson Creek	SW ¼ Sec. 8, T.9N, R.4E. Elk Rock, Washington 15' Quad	535–730	15	306° NW	North-facing slope parallel to blast. Down-timber zone.
Marratta Creek	SW ¼ Sec. 3, T.9N, R.4E. Elk Rock, Cougar, Washington 15' Quad	670–730	13.5	319° NW	SE-facing slope. Direct blast exposure. Tree removal zone.
Castle Creek Ridge	Sec. 23 and 24, T.9N, R.4E. Elk Rock, Cougar, Washington 15' Quad	850–1200	7	312° NW	Ridge running NW parallel to blast. Down-timber zone.
Deer Creek	Sec. 32, T.10N, R.3E. Elk Rock, Washington 15' Quad	365–390	24	300° NW	NE-facing slope nearly parallel to blast. Scorch zone/down-timber zone.

TABLE 9.2
Living Higher Plants Collected (C) or Observed (O) September, 1980, on Five Study Sites, North Fork Toutle Drainage within the Devastated Area Northwest of Mount St. Helens[a]

Species	Post-eruption habitat				Pre-eruption habitat			Life form							Site name					Collected or observed
	Standing dead zone	Blowdown	Debris flow	Levee	Old-growth forest	Young forest	Recent cut/disturbed	Phanerophyte/sprout	Hemicryptophyte/rosette	Geophyte/rhizomes	Annual	Biennial/winter annual	Perennial	Posteruption seedling	Bear	Jackson	Castle	Maratta	Deer	
Equisetum sp., horsetail	X	X			X		X			X			X		X	X				O
Adiantum pedatum L., maidenhair fern (rock outcrop)	X				X					X			X		X	X				C
Athyrium filix-femina (L.) Roth., lady fern	X	X			X	X	X			X			X		X	X				C
Dryopteris austriaca (Jacq.) Woynar, mountain wood fern	X				X	X	X			X			X		X	X				C
Polystichum munitum var. *munitum* (Kaulf.) Presl, sword fern	X	X			X	X	X		X				X		X	X		X	X	C
Pteridium aquilinum (L.) Kuhn., bracken fern	X				X	X	X			X			X		X	X	X			C
Salix sp. L., willow (seedlings on levee, sprouts elsewhere)			X	X		X	X	X					X	X		X	X			O
Alnus sinuata (Regel) Rydb, Sitka alder		X				X		X					X	X			X			O
Rumex sp. L., dock			X	X					?		?		?	?		X				O
Montia sp. L., montia			X	X				?	?		?		?	?		X				O

Species	Abundance
Achlys triphylla (Smith) D.C., deerfoot	C
Berberis nervosa Pursch., cascade oregon grape	C
Dicentra formosa (Andr.) Walp. pacific bleeding heart	C
Mitella sp. L., bishops cap	C
Holodiscus discolor (Pursch) Maxim, ocean spray	C
Rosa sp. L., rose	C
Rubus spectabilis Pursch., salmonberry	O
Rubus parviflorus Nutt., thimbleberry	O
Rubus ursinus Cham. & Schlect., pacific blackberry	O
Rubus laciniatus Willd., evergreen blackberry	O
Lupinus sp. L., lupine	C
Trifolium wormskjoldii Lehm., springbank clover	C
Oxalis oregana Nutt., wood sorrel	O
Acer macrophyllum Pursch., bigleaf maple	C
Acer circinatum Pursch., vine maple	O
Viola sempervirens Greene, evergreen violet	O
Viola purpurea (Kell.) var. *venosa* (Wats.) Brain., purplish violet	C
Circaea alpina L., enchanter's nightshade	C
Epilobium angustifolium L., fireweed	O
Gaultheria shallon Pursh., salal	O
Vaccinium ovatum L., evergreen huckleberry	O
Vaccinium parvifolium Smith, red bilberry	O
Stachys mexicana Benth, mexican betony	C
Mimulus moschatus Dougl., monkey flower	C

TABLE 9.2

LIVING HIGHER PLANTS COLLECTED (C) OR OBSERVED (O) SEPTEMBER, 1980, ON FIVE STUDY SITES, NORTH FORK TOUTLE DRAINAGE WITHIN THE DEVASTATED AREA NORTHWEST OF MOUNT ST. HELENS[a] (continued)

	Post-eruption habitat				Pre-eruption habitat			Life form							Site name					Collected or observed
	Standing dead zone	Blowdown	Debris flow	Levee	Old-growth forest	Young forest	Recent cut/disturbed	Phanerophyte/sprout	Hemicryptophyte/rosette	Geophyte/rhizomes	Annual	Biennial/winter annual	Perennial	Posteruption seedling	Bear	Jackson	Castle	Maratta	Deer	
Penstemon eriantherus Pursh., crested-tongue penstemon (rock outcrop)	X				X			X					X		X					C
Sambucus sp. L., elderberry	X	X		X	X	X		X					X		X	X	X		X	O
Adenocaulon bicolor Hook., trail plant	X	X				X	X			X					X	X				C
Anaphalis margaritacea (L.) B. & H., pearly everlasting	X	X				X	X			X			X		X	X	X	X		C

Species									Life form[a]
Cirsium sp. Mill., thistle (seedlings, <1 week of age)	X	X		X	X		X	X X X X	O
Cirsium undulatum (Nutt.) Spreng., wavy leaved thistle	X	X		X	X		X	X	C
Hieracium pilosella L., mouse ear hawkweed	X			X	X			X	C
Hypochaeris sp. L., cats ear	X X			X	X X		X	X	C
Petasites frigidus (L.) Fries var. *palmatus* (Ait.) Cronq., coltsfoot	X	X		X	X		X	X X	O
Senecio sylvaticus L., wood groundsel	X X	X X		X X	X X		X	X X X	C
Gramineae spp.	X X	X X		X X	X		X	X X X	O
Agrostis exarata Trin. ssp. *minor* (Hook.) Hitchc., spike bentgrass	X X			X	X X		X	X	C
Bromus orcuttianus Vasey, orcutt brome	X			X	X		X	X	C
Festuca arundinacea Schreb., reed fescue	X X			X X	X X		X	X X	C
Holcus lanatus L., velvetgrass	X			X	X		X	X	C
Lysichitum americanum Hulten & St. John, skunk cabbage	X	X			X		X		O
Streptopus sp. Michx., twisted stalk	X		X	X	X		X	X	O

[a]Post= 1 and preeruption habitat occurrence and life forms are indicated. A question mark indicates a tentative assignment

10

How Plants Survive Burial:
A Review and Initial Responses
to Tephra from Mount St. Helens

Joseph A. Antos and Donald B. Zobel

ABSTRACT

Literature reporting responses of plants buried by volcanic tephra, sand dunes, dust storms, and stream alluvium suggests a generalized model for recovery of plants from burial. The model relates the potential for plant survival to plant characteristics, such as deciduousness and amount of shoot elongation and adventitious rooting, and to environmental factors, such as the depth and timing of the deposit. Observations during the first year after the eruption of Mount St. Helens indicate the importance of the season during which the volcanic deposits were formed and the factors that modify the extent of burial. Snowpack at the time of tephra deposition greatly decreased the survival of woody plants. Microsites next to logs and along erosion channels were of great importance to the survival of some plant species. Many plants that remained buried at the end of the first summer were alive and had shoots growing into the overlying layer of tephra. Adventitious roots in the tephra were common on shrubs, except for the ericaceous species that usually dominate the forest understory. During the first season after the eruption, almost no seedlings appeared on the tephra.

INTRODUCTION

The May 18, 1980, eruption of Mount St. Helens produced a wide array of impacts on ecosystems. These ranged from total destruction near the volcano to a minor accumulation of tephra (volcanic aerial ejecta) farther away. Although the devastation near the mountain is most impressive, an even larger area beyond the range of the blast was affected by significant tephra deposition. A major concern is how forest understory plants respond to tephra deposition in areas where trees are left intact. The ultimate goal is to understand how plants morphologically respond to burial and how their survival is related to their normal growth form.

Tephra deposition is probably the most extensive type of volcanic influence on ecosystems. In some regions, including the area around Mount St. Helens, tephra has been deposited repeatedly, influencing vegetation composition and species evolution. Tephra deposition can affect a plant both mechanically and physiologically. It can bury small plants and form a physical barrier to the penetration of stems. In addition, the condition of the soil beneath the tephra can be altered significantly. To survive, a plant must not only grow through the deposit but also cope with the altered soil environment.

Burial of plants is not unique to volcanic eruptions; sand dune movement, dust storms, and alluvial deposition also bury plants. These various depositional situations share some similarities with tephra deposition, but they also differ in some ways. The objectives of this report are to: (1) summarize information about other depositional environments and compare them to tephra fall, (2) summarize information on plant responses to tephra deposition, (3) present a model of plant response to burial, and (4) report first-year observations from Mount St. Helens and relate them to the model.

BURIAL BY DUNE SAND, DUST, AND ALLUVIUM

In active sand dunes, movement of the substrate is chronic, and many plant species are eliminated by burial (van der Valk 1974). The predictable nature of burial in sand dunes has permitted plants to adapt to it. Some grasses that grow in areas of sand accretion not only tolerate but require deposition to retain their maximum vigor (Au 1974, Godfrey and Godfrey 1976, Hope-Simpson and Jefferies 1966, Laing 1958, Marshall 1965, Martin 1959, Wagner 1964). In some species old roots degenerate and new adventitious roots appear on the part of the plant in the new sand deposit. Sand dune species tend to produce vigorous rhizomes, adventitious roots, and large seeds (Godfrey and Godfrey 1976, Holton and Johnson 1979, Kumler 1963). Sand deposition can inhibit plant establishment by burying seed and seedlings (Holton and Johnson 1979, Laing 1958, van der Valk 1974, Wagner 1964). Shoots of seedlings from seeds with large food reserves, however, can penetrate deeper deposits (Holton and Johnson 1979, Kumler 1963, van der Valk 1974), thus selection for large seeds can occur in the sand dune environment.

During the mid-1930s, wind erosion deposited dust over significant areas of the Great Plains, destroying vegetation (Weaver and Albertson 1936, 1940). Deposits of 1.2–3.8 cm of dust killed most short grasses and produced an ideal habitat for weed invasion. Buffalo grass (*Buchloe dactyloides*) grew well only where dust deposits were <0.8 cm deep (Weaver and Albertson 1936), whereas blue grama (*Bouteloua gracilis*) survived dust 5 cm deep by forming vertical rhizomes and new crowns (Robertson 1939). *Agropyron smithii*, which could penetrate deposits as deep as 26 cm and could spread rapidly by rhizomes, soon dominated areas covered with dust (Mueller 1941, Robertson 1939, Weaver and Albertson 1936). Big bluestem (*Andropogon gerardi*), although a tall grass, was destroyed by dust 2.5–7 cm deep (Robertson 1939, Weaver and Albertson 1936). In contrast, this species has been observed to penetrate 15–20 cm of loose soil deposited by pocket gophers (Weaver and Fitzpatrick 1934), indicating that the nature of the deposit can be very important in determining survival.

An excellent experimental study concerning plant response to burial is Mueller's (1941) research on dust deposition. Using established clumps of prairie plants, she placed dust deposits of 7.5, 15, and 30 cm on experimental plots and recorded the number of emergent shoots after 1 yr of burial. Only one of eight species tested—the vigorously rhizomatous *Agropyron smithii*—grew through 30 cm of dust deposit, and the 7.5 and 15 cm deposits greatly reduced the number of shoots of most species. *Andropogon gerardi*, with slow-growing rhizomes 2.5–5.0 cm deep in the soil, failed to penetrate 15 of dust. The tall grasses *Calamovilfa longifolia*, found on stabilized sands, and *Spartina pectinata*, which is dominant in sloughs, grew well where deposits were 15 cm or less but failed to appear above 30 cm of dust. *Solidago glaberrima* adjusted rapidly to 15 cm of dust by producing vertical rhizomes; the second season, the vertical rhizomes produced shoots and new horizontal rhizomes in the dust at the normal depth of 2.5–7.5 cm. Rhizomes of most species grew slower than 30 cm/yr, such that if the rhizomes did turn up in deep deposits, they would not be able to reach their normal depth below the surface in 1 yr. Mueller's study illustrated that the growth rate of the rhizome and its ability to turn upward are important for the survival of grasses after burial. The data indicated that there is no direct relationship between the normal size of a plant and the depth of deposit that it can penetrate. Aerial shoots of large, robust plants may be unable to penetrate deposits only a small percentage of their normal height.

Alluvial deposition is important in flood-plain morphology and resulting vegetation composition (Johnson et al. 1976, Lindsey et al. 1961, Wistendahl 1958); however, its effects on established plants are apparently often secondary to those of the concomitant flooding and thus are rarely studied. Redwood (*Sequoia sempervirens*) can tolerate alluvial deposition while competing trees are often destroyed (Stone and Vasey 1968). Redwood roots penetrate the new deposit from below, and eventually adventitious roots form a totally new root system in the alluvium.

Deposition of loess was a widespread disturbance during periods of major glacial activity (Curtis 1959), but its effects on plants have apparently not been investigated.

TEPHRA DEPOSITION

The effects of tephra on vegetation involve (1) damage caused directly or indirectly by the tephra and (2) subsequent recovery from such damage by the survival and expansion of previously existing plants or establishment of new individuals. Damage to plants can result from chemical or mechanical factors. Although direct effects on leaf surfaces by tephra from Mount St. Helens did occur in 1980 (Mack 1981, Hinckley et al. 1984), this chapter primarily considers mechanical problems resulting from tephra deposits on the ground.

The most relevant study of the effects of tephra on vegetation was by Griggs (1918, 1919*a,b*, 1922) following the 1912 eruption of Katmai Volcano in Alaska. An impressive recovery of herbs occurred despite the 25–30 cm of tephra deposited on Kodiak Island. Many species, especially grasses, grew through the tephra to form dense stands (Griggs 1918), and robust grasses such as *Deschampsia cespitosa* penetrated deposits 50 cm deep. Where drifts developed, however, plants that did not produce new roots in the tephra eventually died. In the Sitka spruce (*Picea sitchensis*) forest, understory herbs did not generally appear during the first year. *Rubus pedatus*, an abundant understory species before the eruption, did not appear until the third year after the eruption, but was common by the fourth year (Griggs 1918). Other forest herbs showed similar patterns of recovery. *Equisteum arvense* was the most successful herb (Griggs 1918, 1919*b*), colonizing large areas of the tephra with its extensive rhizome systems. *Equisetum* penetrated as much as 1 m of tephra, more than any other herbaceous species. *Elymus arenarius*, which was very successful in tephra near the seashore, is strongly rhizomatous and well adapted to withstand sand burial (Griggs 1919*b*).

After the eruption of Katmai, erosion was of great importance to vegetation recovery (Griggs 1918, 1919b). This eruption covered many square kilometers with tephra so deep that only trees and large shrubs protruded above it (Griggs 1918, 1922); after about 5 yr, the only significant herbaceous cover that had developed was in areas freed of tephra by erosion. Some plants sprouted when uncovered after 3 yr of burial (Griggs 1919b), indicating a remarkable ability to remain dormant but alive under the tephra. Most species withstood burial of 1 yr or more. Griggs (1919b) felt that the species composition of vegetation on areas freed of tephra closely resembled that before burial; thus species survival in the region was related more to the nature of the sites exposed than to specific tolerance characteristics of the plants. Although Kodiak Island received 25–30 cm of tephra during the eruption, erosion removed much of the deposit in a few years (Griggs 1918). Common cushion plants in ridgetop tundra were unable to grow through deep tephra, but erosion soon removed most of the deposit and many plants survived. On grassy tundra hillsides, many species penetrated the tephra, but erosion greatly aided vegetation recovery. In contrast, erosion deposited tephra on low, boggy sites where few plants survived burial.

The importance of erosion to vegetation recovery is not unique to the Katmai eruption. Two generalizations can be drawn from work in Mexico (Eggler 1959, 1963), St. Vincent (Beard 1976), the Philippines (Brown et al. 1917, Gates 1914), and New Guinea (Taylor 1957): (1) tephra erodes very rapidly in wet climates, and (2) recovery of preexisting plants is facilitated where erosion occurs. The nature of the tephra can be important; apparently pumice does not erode as easily as volcanic ash, possibly because of its high permeability and large particle size.

Tephra deposits tend to persist on gently sloping or flat topography, making the structural characteristics of plants more important for survival than on steep slopes where much of the tephra can be removed by erosion. Trees have survived more than 2 m of burial in Mexico (Eggler 1948), Alaska (Griggs 1918, 1919a,b), Hawaii (Smathers and Mueller-Dombois 1974), and the Philippines (Brown et al. 1917), and the grass *Saccharum spontaneum* has survived as much as 1.5 m of burial in New Guinea (Taylor 1957). Survival in >30 cm of tephra is reported for herbaceous species in many areas. Grasses with a "running habit" survived burial better than others in the Philippines (Brown et al.1917). Many of the surviving plants sent adventitious roots into the tephra (Brown et al. 1917, Eggler 1948, Griggs 1918, 1919a,b, Smathers and Mueller-Dombois 1974). The root system of surviving pines grew up into the tephra near Paricutin Volcano in Mexico (Eggler 1959, 1963). Although these pines grew well, they developed abnormal wood (Eggler 1967). Spruce in Alaska (Griggs 1922) grew faster following tephra deposition, possibly because of reduced competition or some other favorable environmental modification.

The establishment of seedlings had usually not been as important to vegetation recovery following tephra deposition as the survival of previously established plants (Eggler 1959, 1963; Griggs 1918, 1922). Seedling establishment in pure tephra is uncommon and appears to be inhibited by a nitrogen deficiency (Brown et al. 1917; Eggler 1959; Griggs 1919b, 1933; Shipley 1919). Erosion, tephra accumulation, hard surface crusts, low water holding capacity, and acidity also have been implicated (Brown et al. 1917, Eggler 1959; Griggs 1919b). Where tephra becomes mixed with old soil or organic matter, seedlings establish more readily. On Kodiak Island, seedlings of all major species in the *Picea sitchensis* forest appeared, at least in small numbers, during the fourth season (Griggs 1918).

These studies of vegetation recovery following tephra deposition suggest several important generalizations: (1) recovery of previously established plants is more important than seedling establishment; (2) erosion greatly facilitates plant recovery; (3) some exhumed plants can still recover after 3 yr or more of burial; (4) herbaceous plants can take as long as 4 yr to penetrate the deposit; (5) herbaceous growth form is related to ability to survive; (6) woody plants can survive in deeper deposits than herbaceous plants; (7) growth of woody plants can sometimes increase after tephra deposition; (8) surviving plants usually produce adventitious roots in the tephra; and (9) because the concentration of available nutrients is often low in tephra, species that normally colonize nutrient-poor substrates should have an advantage when becoming established on tephra.

A MODEL OF PLANT RECOVERY

From these generalizations, a model of how plants survive burial has been developed (Fig. 10.1). Following tephra deposition, the plant can exist in one of two states: (1) either it is totally buried or (2) has some emergent shoots (Fig. 10.1, path a). If the plant is a deciduous herb, emergent shoots are only a temporary success; once the shoots die back, the plant will be buried (Fig. 10.1, path b). Usually the shoot dies back to or below the old soil surface, but it is possible that the bases of

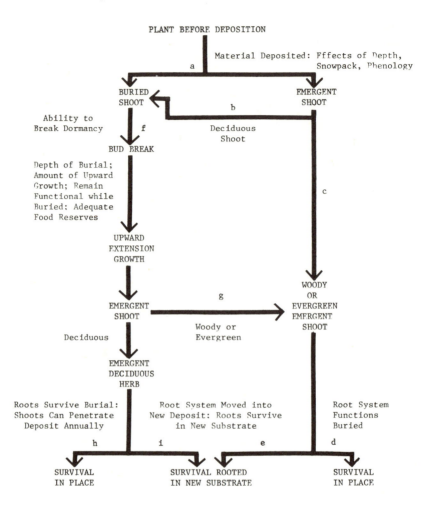

FIG. 10.1   Responses necessary for plant survival of burial. See text for explanation. The term *roots* here implies all portions of the plant that normally grow underground.

normally deciduous shoots would not die completely after burial. In this case, being initially emergent could give the plant a structural advantage for future penetration of the tephra. If the emergent shoots are perennial, the plant is well on the way to survival (Fig. 10.1, path c). The crucial factor at this point may be whether the root system can function in buried soil. If it can, the plant will succeed (Fig. 10.1, path d); if it cannot, success can still occur if the underground structures can develop in the new material (Fig. 10.1, path e). An intermediate situation can occur in which the old root system functions in an impaired condition while new roots develop in the deposit.

Buried plants must complete several processes before aerial shoots can escape through a deposit (Fig. 10.1, path f). First, bud break must occur. Once a shoot has begun to grow, emergence depends on the amount and direction of growth, the depth of the deposit, the force necessary to penetrate the deposit, the ability of the shoot to function while buried, and the adequacy of food reserves. A plant may emerge through cracked or thin deposits when penetration of normal tephra is not possible. If the emerging shoot is woody or evergreen (Fig. 10.1, path g), it need only retain its root system in the old soil or form a new system in the deposit to succeed. If the emergent shoot is deciduous, two mechanisms for survival exist. First, the root system and perennating structures can remain in the old soil and the shoots penetrate the deposit annually (Fig. 10.1, path h). This requires that the plant remains functional in the old soil, is able to provide indefinitely the extra energy required for penetration, and can tolerate any phenological delay caused by penetration of the deposit. This approach can be used by plants in shallow deposits or by large, robust herbaceous plants in deeper deposits. Second, small herbaceous plants, or all herbaceous plants in deep deposits, may have little alternative to moving their underground structures (roots, perennating organs, or both) into the deposit. Survival in this manner (Fig. 10.1, path i) requires an ability to tolerate the microenvironment there. Once the perennating structure of a plant has reached its normal depth, it has escaped the structural difficulties imposed by the new deposit. No matter how these difficulties are overcome, survival can occur only if the plant can maintain a functional root system in the modified environment of the old soil, in the new conditions in the deposit, or in some combination of the two.

In addition to their similarities, there are many important differences among environments where plants are buried (Table 10.1) that affect the mechanisms of survival presented in Figure 10.1. Sand dunes and dust storms normally deposit material similar to that in which the plants were already growing. Thus colonization of the new material is not likely to be limited by nutrient shortages, chemical toxicity, or water deficiencies greater than in the buried soil. Stream alluvium is more apt to vary in texture, but it is still likely to be similar to the buried soil. In contrast, tephra is often very different from the underlying soil and is usually less favorable for root growth. Tephra is often low in available nutrients, can have poor water-holding capacity, and can contain toxic chemicals (Working Group on the Damage to Forest Trees by the Ejecta 1965). The thermal characteristics of pumice result in extreme soil surface temperatures (Cochran et al. 1967). In addition, mycorrhizal inoculum is more apt to be lacking in tephra than in other types of deposits. The net result is that movement of root and rhizome systems into the tephra may be less extensive than in other types of deposits.

The frequency of deposition will vary greatly, from continuous for some sand dunes, to infrequent and erratic for tephra deposition. The disturbance with the greatest frequency should provide the best opportunity for plant adaptation. Tephra covers large, continuous areas at a single time and thus allows little opportunity for spatial escape. Other depositional environments provide a finer-grained mosaic of disturbed and undisturbed sites, increasing the chance that members of a species may escape to recolonize the disturbed areas from within.

FIRST-YEAR RESPONSES TO MOUNT ST. HELENS TEPHRA

During the summer of 1980, an extensive reconnaissance was conducted northeast of Mount St. Helens in the area that was affected only by tephra fall. Six intensive study sites were chosen (Fig. 10.2) along a gradient of tephra depth ranging from 2 to18 cm, all in old-growth forest on relatively gently sloping topography. Established at each site were 150 plots, each 1 m². We cleared 50 plots of all tephra to obtain an idea of which plants had been buried and to provide an immediate "erosion" treatment for comparison with the 100 plots that were covered with natural tephra. Cover and density of all vascular plants and cover of mosses in each plot were recorded during August and September, 1980. All plots were photographed, plants were mapped on some plots, and detailed observations were made of the nature of the tephra deposit. The plots were permanently marked to be used for reexamination during the following several summers.

From the plot data and field observations, the importance of a number of factors affecting plant survival and vegetation recovery can be assessed. One very important factor modifying plant burial was snowpack. On May 18, 1980, the snowpack on the sites was only partly melted. Of two sites each at locations 1 and 3 (Fig. 10.2), one had snow at the time of the eruption while the other did not. This allowed a comparison of plant response to tephra with and without a snowpack. The presence of snow when the tephra was deposited greatly reduced

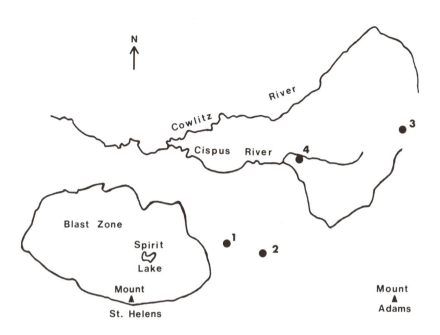

FIG. 10.2   Study site locations numbered from 1 (deepest tephra) to 4 (shallowest tephra). Locations 1 and 3 have two intensive study sites each: one that was snow covered at the time of the May 18, 1980, eruption and one that was mostly snow free.

the cover and density of woody understory plants (Antos and Zobel 1982). The ability of the snowpack to flatten plants against the ground is responsible for this effect. Where snow was absent, tephra fell through the erect leafless shrubs, burying only those stems that were shorter than the tephra was deep. Where snow was present, however, those stems buried beneath the snow were also completely below the tephra. As the snow melted, they were trapped beneath the tephra. Excavation of the tephra showed that shrubs normally 1.5-m tall were completely buried by as little as 4.5 cm of tephra deposited on snow, and conifers 3-m tall were largely buried by 18 cm of tephra deposited on snow. In contrast to this result, snowpack protected small trees and shrubs in areas devastated by the lateral blast during the May 18 eruption (see Chapter 6, this volume).

    In contrast to other depositional mechanisms, volcanism is independent of season and weather; tephra fall is equally likely at any time of the year. The plant response, however, is greatly modified by the timing of the event during the year. The effect of snow emphasizes the importance of the timing of tephra fall. These timing considerations are critically important at the first step in the burial model (Fig. 10.1, path a). Even if a plant is relatively large compared to the depth of the deposit, it can still become buried if it is trapped beneath snow. Snow is not the only reason that timing is important. The phenological stage of the plant can drastically affect the impact of tephra. Shrubs that had leafed out by the time of the eruption were damaged by ash accumulations on the foliage and possibly by some transient chemical toxins, while shrubs that were still leafless escaped such injury; similar damage was seen in herbs.

If woody plants escape burial, the ability to form adventitious roots may be important for their survival (Fig. 10.1). A major difference was noted in adventitious root production between the ericaceous shrubs, which dominate the forest understory, and species in more open sites (Table 10.2). No ericaceous shrub formed adventitious roots during 1980 (although we have subsequently seen some), whereas other shrub species did so. This may be of great importance to long-term species survival.

The first season is too early to make a detailed analysis of how growth form relates to survival, but a few conclusions about initial plant responses can be reached by comparing cleared plots with plots covered by natural tephra (Table 10.3). Tree seedlings and shrubs showed moderate reductions in abundance, primarily because of snow, in both the deep and shallow tephra. Evergreen subshrubs were greatly reduced by deep tephra, but only slightly by shallow tephra deposits. Deciduous, herbaceous plants probably had a similar pattern. The data, however, were biased because most of the delicate foliage was destroyed during clearing of the tephra. As expected, mosses were almost totally covered even by very thin deposits of tephra. In the deep tephra, larger woody plants had an advantage, but in the thinner deposits, herbaceous plants and subshrubs maintained their abundance about equally. The size advantage of woody plants was offset by burial under snowpack and by their limited ability to penetrate the deposit once buried. By the end of summer, buried conifer branches were nearly dead, and buried shrubs had only small, weak shoots in the tephra. Many buried shrub stems were dying back from the tips, and some live ones had failed to break bud. In contrast, some herbaceous plants had penetrated the deepest deposits.

At least some herbaceous plants have the ability to survive prolonged burial; this affects path f of the model (Fig. 10.1). Excavation revealed that there were large numbers of herbaceous shoots growing upward into the tephra. Many of these may eventually reach the surface a year or more after burial. Consequently, it is inappropriate to predict eventual survival of herbaceous plants from first-year emergence. These observations are consistent with those of Griggs (1918, 1922) on Kodiak Island.

Microsites with unusually thin or absent tephra deposits are of extreme importance for the survival of many plant species. Few if any species have been eliminated from the study sites, mainly because of microsite refugia. Logs, tree bases, rocks, steep slopes, and erosion channels all provide areas where tephra was thin or absent. Many species of moss survived on the sides of or under large logs, while being completely covered with tephra elsewhere. Large logs may be a critical refugium for the survival and eventual recolonization of such species. Many herbaceous plants emerged primarily in microsites covered by thin tephra. Erosion has already been important in providing microsite refugia and may eventually modify the entire nature of the deposit, especially on the steeper slopes. Cracks caused by the shrinkage of the surface crust or by large plants emerging from the tephra also provided escape routes for many individuals of species that would normally not have penetrated the intact deposit. In contrast, some herbaceous plants that elongated sufficiently to escape were trapped beneath the hard crust that developed during 1980 when the fine ash on top of the deposit was wetted and then dried.

Seedlings were almost nonexistent during the first season. In September 1980, a few *Tsuga heterophylla* seedlings germinated from newly released seed of the current year's crop. The lack of seedlings probably resulted from both an absence of seed on the tephra and adverse germination conditions. The finding that early vegetation recovery was predominantly from previously established plants, rather than from seedling establishment, is consistent with results from studies of other volcanic eruptions.

ACKNOWLEDGMENTS

We thank Tom Hill, Charlie Halpern, and Brad Smith for assistance with fieldwork. The field research was funded by a grant from the National Science Foundation.

LITERATURE CITED

ANTOS, J. A., and D. B. ZOBEL. 1982. Snowpack modification of volcanic tephra effects on forest understory plants near Mount St. Helens. *Ecology* 63:1969–1972.

AU, S.-F. 1974. Vegetation and ecological processes on Shackleford Bank, North Carolina. *Natl. Park Serv. Sci. Monogr. Series 6.* 86 pp.

BEARD, J. S. 1976. The progress of plant succession on the Soufriere of St. Vincent: Observations in 1972. *Vegetatio* 31:69–77.

BROWN, W. H., E. D. MERRILL, and H. S. YATES. 1917. Revegetation of Volcano Island, Luzon, Philippine Islands, since eruption of Taal Volcano in 1911. *Phil. J. Sci. Sect. C.* 12(4):177–248.

COCHRAN, P. H., L. BOERSMA, and C. T. YOUNGBERG. 1967. Thermal properties of a pumice soil. *Soil Sci. Soc. Amer. Proc.* 31:454–459.

CURTIS, J. T. 1959. *The Vegetation of Wisconsin.* Univ. of Wisconsin Press, Madison. 657 pp.

EGGLER, W. A. 1948. Plant communities in the vicinity of the volcano El Paricutin, Mexico, after two and a half years of eruption. *Ecology* 29:415–436.

———. 1959. Manner of invasion of volcanic deposits by plants with further evidence from Paricutin and Jorullo. *Ecol. Monogr.* 29:267–284.

———. 1963. Plant life of Paricutin Volcano, Mexico, eight years after activity ceased. *Amer. Midl. Natur.* 69:38–68.

———. 1967. Influence of volcanic eruptions on xylem growth patterns. *Ecology* 48:644–647.

GATES, F. C. 1914. The pioneer vegetation of Taal Volcano. *Philippine J. Sci. Sect. C.* 9:391–434.

GODFREY, P. J., and M. M. GODFREY. 1976. Barrier island ecology of Cape Lookout National Seashore and vicinity, North Carolina. *Natl. Park Serv. Sci. Monogr. Series 9.* 160 pp.

GRIGGS, R. F. 1918. The recovery of vegetation at Kodiak. *Ohio J. Sci.* 19:1–57.

———. 1919a. The character of the eruption as indicated by its effects on nearby vegetation. *Ohio J. Sci.* 19:173–209.

———. 1919b. The beginnings of revegetation in Katmai Valley. *Ohio J. Sci.* 19:318–342.

———. 1922. *The Valley of Ten Thousand Smokes.* Nat. Geog. Soc., Washington, D.C. 340 pp.

———. 1933. The colonization of the Katmai ash, a new and inorganic "soil." *Amer. J. Bot.* 20:92–113.

HINCKLEY, T. M., H. IMOTO, K. LEE, S. LACKER, Y. MURIKAWA, K. A. VOGT, C. C. GRIER, M. R. KEYES, R. O. TESKEY, and V. SEYMOUR. 1984. Impact of tephra deposition on growth in conifers: The year of the eruption. *Can. J. For. Res.* 14:731–739.

HOLTON, B., JR., and A. F. JOHNSON. 1979. Dune scrub communities and their correlation with environmental factors at Point Reyes National Seashore, California. *J. Biogeogr.* 6:317–328.

HOPE-SIMPSON, J. F., and R. L. JEFFERIES. 1966. Observations relating to vigour and debility in marram grass (*Ammophila arenaria* [L] Link). *J. Ecol.* 54:271–274.

JOHNSON, W. C., R. L. BURGESS, and W. R. KEAMMEROR. 1976. Forest overstory vegetation and environment on the Missouri River floodplain in North Dakota. *Ecol. Monogr.* 46:59–84.

KUMLER, M. L. 1963. Succession and certain adaptive features of plants native to the sand dunes of the Oregon coast. Ph.D Thesis, Oregon State Univ., Corvallis. 149 pp.

LAING, C. C. 1958. Studies in the ecology of *Ammophila breviligulata*: Seedling survival and its relation to population increase and dispersal. *Bot. Gaz.* 119:208–216.

LINDSEY, A. A., R. O. PETTY, D. K. STERLING, and W. VAN ASDALL. 1961. Vegetation and environment along the Wabash and Tippecanoe Rivers. *Ecol. Monogr.* 31:105–156.

MACK, R. N. 1981. Initial effects of ashfall from Mount St. Helens on vegetation in eastern Washington and adjacent Idaho. *Science* 213:537–539.

MARSHALL, J. K. 1965. *Corynephorus canescens* (L.) D. Beauv. as a model for the *Ammophila* problem. *J. Ecol.* 53:447–463.

MARTIN, W. E. 1959. The vegetation of Island Beach State Park, New Jersey. *Ecol. Monogr.* 29:1–46.

MUELLER, I. M. 1941. An experimental study of rhizomes of certain prairie plants. *Ecol. Monogr.* 11:165–188.

ROBERTSON, J. H. 1939. A quantitative study of true prairie vegetation after three years of extreme drought. *Ecol. Monogr.* 9:431–492.

SEYMOUR, V. A., T. M. HINCKLEY, Y. MORIKAWA, and J. F. FRANKLIN. 1983. Foliage damage in coniferous trees following volcanic ashfall from Mt. St. Helens. *Oecologia* 59:339–343.

SHIPLEY, J. W. 1919. The nitrogen content of volcanic ash in the Katmai eruption of 1912. *Ohio J. Sci.* 19:213–223.

SMATHERS, G. A., and D. MUELLER-DOMBOIS. 1974. Invasion and recovery of vegetation after a volcanic eruption in Hawaii. *Natl. Park Serv. Sci. Monogr. Ser. 5.* 129 pp.

STONE, E. C., and R. B. VASEY. 1968. Preservation of coast redwood on alluvial flats. *Science* 159:157–161.

TAYLOR, B. W. 1957. Plant succession on recent volcanoes in Papua. *J. Ecol.* 45:233–243.

VAN DER VALK, A. G. 1974. Environmental factors controlling the distribution of forbs on coastal foredunes in Cape Hatteras National Seashore. *Can. J. Bot.* 52:1057–1073.

WAGNER, R. H. 1964. The ecology of *Uniola paniculata* L. in the dune-strand habitat of North Carolina. *Ecol. Monogr.* 34:79–96.

WEAVER, J. E., and F. W. ALBERTSON. 1936. Effects of the great drought on the prairies of Iowa, Nebraska, and Kansas. *Ecology* 17:567–639.

———. 1940. Deterioration of midwestern ranges. *Ecology* 21:216–236.

WEAVER, J. E., and T. J. FITZPATRICK. 1934. The prairie. *Ecol. Monogr.* 4:109–296.

WISTENDAHL, W. A. 1958. The floodplain of the Raritan River, New Jersey. *Ecol. Monogr.* 28:129–153.

WORKING GROUP ON THE DAMAGE OF FOREST TREES BY THE EJECTA. 1965. Damage to forest trees by the Shinmoe-dake ejecta of the Kirishima Volcano Cluster, following eruption in 1959. *Bull. Gov. Forest Exp. Station, No. 182.*, pp. 67–112.

TABLE 10.1

A COMPARISON OF CHARACTERISTICS OF VARIOUS DEPOSITIONAL
ENVIRONMENTS IMPORTANT TO PLANT SURVIVAL OF BURIAL

	Tephra	Sand dunes	Dust	Alluvium
Chemical toxicity	Sometimes	Unlikely	Unlikely	Unlikely
Nutrient deficiency	Likely	Likely	Unlikely	Unlikely
Frequency	Low	Highest	Low	Occasional
Predictability	Low	High	Low	Moderate
Relative potential for plant adaptation	Low	High	Low	Moderate
Estimated abundance of refugia within the deposit	Low	High	High	High
Correlation with season and weather	None	Moderate	High	High

TABLE 10.2

ADVENTITIOUS ROOT PRODUCTION BY SHRUBS DURING THE FIRST
SUMMER AFTER TEPHRA DEPOSITION (1980)

Species	Habitat	Adventitious roots[a]
Menziesia ferruginea	Forest	0
Rhododendron albiflorum	Forest	0
Vaccinium alaskense	Forest	0
V. membranaceum	Forest	0
V. ovalifolium	Forest	0
Ribes lacustre	Open	+
Rubus parviflorus	Open	+
R. spectabalis	Open	+
Alnus sinuata	Open	++
Salix sp.	Open	++
Spiraea densiflora	Open	++

[a]0 = none found; + = common; ++ = abundant.

TABLE 10.3
COVER (%) AND DENSITY (NUMBER/m²) BY GROWTH FORM FOR
DEEP AND SHALLOW TEPHRA LOCATIONS[a]

| | Tephra depth[b] | | | |
| | 15-18 cm | | 4.5 cm | |
Plot type	NT	C	NT	C
Tree seedlings				
Cover	3.0	8.8	7.4	7.4
Density	0.2	9.0	0.8	5.2
Shrubs				
Cover	7.6	15.0	9.4	27.8
Density	1.8	4.2	2.5	4.0
Evergreen subshrubs				
Cover	0.02	1.2	0.8	1.0
Density	0.05	10.9	2.4	3.9
Beargrass				
Cover	0	0	5.4	9.4
Density	0	0	1.3	1.5
Deciduous herbs				
Cover	0.03	0.3	4.2	1.6
Density	0.15	6.6	5.8	3.7
Mosses				
Cover	0.06	18.2	0.05	21.1

[a]Values from two stands are averaged at each location.

[b]NT = natural tephra; C = cleared plots.

11

Effects of Mount St. Helens Ashfall in Steppe Communities of Eastern Washington: One Year Later

Richard N. Mack

ABSTRACT

Effects of the May 18, 1980, ashfall from Mount St. Helens on steppe vegetation have been studied during the year since the summer of 1980. The ashfall provided the opportunity for researchers to decipher the record of vegetation changes that occurred in response to earlier ashfalls, such as that from Mount Mazama, and to examine these grassland communities after perturbation. Detectable ash damage to vegetation was generally limited to areas where silt-sized ash accumulations were >1 cm thick. Plant damage (both through mechanical overloading and later necrosis) was most extensive in the summer of 1980 among species with an acaulescent (e.g., *Balsamorhiza careyana*) or prostrate habit or with broad, sheathing leaves (e.g., *Veratrum californicum*). Both annual and perennial grasses showed negligible damage and continued to form caryopses in the late spring and early summer of 1980. Interruption of flowering through the coating of floral parts by ash had little effect on seed production of most species because additional flowers were produced after May 18. Most of the 1980 seed crop of *Balsamorhiza sagittata* and *V. californicum* were lost when the peduncles collapsed with the ash coating. As the fine-textured ash hardened over summer, the seed bank was split into two components: seeds below the ash (the residual seed bank in the soil) and seeds cast after May 18, 1980. In the *Artemisia tridentata*–dominated communities, seed foraging has been altered as seeds cast after the ashfall are now mainly confined to cracks in the ash or microdepressions in the hardened ash surface. Seedling recruitment in the first year after the ashfall was largely confined to these cracks and microdepressions. Most of the annual dicots emerging in the autumn of 1980 died without producing seeds. In contrast, winter annual grasses consistently produced seeds on the same sites. Demise of cryptogams still buried under the ash could initiate wholesale change in species composition of communities with the entry of alien weeds.

INTRODUCTION

Volcanic ashfalls have been sufficiently frequent worldwide in the past century to allow assessment of these usually short-term perturbations on a variety of ecosystems ranging from tropical montane forests (Smathers and Mueller-Dombois 1974) to boreal forests and heaths (Griggs 1918). However, the May 18, 1980, deposition of ash from Mount St. Helens on the steppe vegetation in eastern Washington is perhaps without parallel in modern times (Fig. 11.1). The apparent rarity of this event results from there being few regions on the earth's surface where physical features include both the steppe environment and active volcanoes. Only in New Zealand (Godley 1975) and in the *Agropyron spicatum* Province (*sensu* Daubenmire 1978) of the interior Pacific Northwest do steppe communities currently occur within the likely range of ashfall (Macdonald 1972).

Before May, 1980, the effects of ash deposition on steppe plants could only be inferred from observations of similar life forms and other vegetation types (e.g., boreal forest and tropical alpine) subjected to ashfall. Such inferences seem to be most unsatisfactory when they are based on the accumulating circumstantial evidence from pollen analyses of effects of earlier Quaternary volcanic ashfalls on these temperate grasslands (Mack et al. 1978, Blinman et al. 1980). The large ashfalls from Mount Mazama (~6700−6900 yr ago) may have caused short-term regional alteration of the physical environment favoring arid land shrubs such as *Artemisia* at the expense of conifers (particularly *Pinus*). At Big Meadow, 750 km northwest of Crater Lake (the site of Mount Mazama), *Artemisia* pollen percentages increase and those of *Pinus* decrease in the first sample level above the Mazama ash layer (Mack et al. 1978) (Fig. 11.2). A similar response is recorded at Tepee Lake in northwestern Montana (Mack et al. 1983). Blinman et al. (1980) ascribe a possible mulching role for Mazama ash, noting the importance of phenology in amplifying any such response.

Once the effect of the May 18 ashfall is quantified, the magnitude of vegetation change coincident with the Mazama ashfall can possibly be gauged. Within the same sedimentary basin, pollen records bracketing Mazama ash can be compared with pollen assemblages formed immediately before and after deposition of the ash from Mount

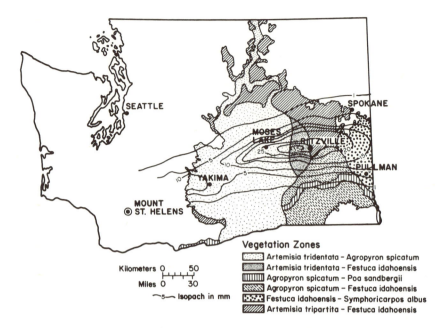

Vegetation Zones

- [:::] Artemisia tridentata – Agropyron spicatum
- [▓▓] Artemisia tridentata – Festuca idahoensis
- [||||||] Agropyron spicatum – Poa sandbergii
- [▨▨▨] Agropyron spicatum – Festuca idahoensis
- [▦▦] Festuca idahoensis – Symphoricarpos albus
- [▨▨▨] Artemisia tripartita – Festuca idahoensis

FIG. 11.1   Distribution and thickness (mm) of ash compacted by rain from the May 18, 1980, eruption of Mount St. Helens (Quinn and Folsom 1980) in relation to the steppe vegetation zones of eastern Washington (Daubenmire 1970). The fan-shaped ashfall area bisected several vegetation zones, allowing direct comparison of possible ash-induced effects on vegetation with control sites on the same habitat types south and/or north of the deposition area. Ash texture generally graded from sand-sized particles at the western margin of the Columbia Basin to silt-sized particles farther east.

St. Helens. Textural and chemical properties of the various Holocene tephras, including those of Mount St. Helens, will also need to be compared since these influence the effects of ash on vegetation. For example, percolation of water in coarse ash (particle size >2 mm) is much more rapid than in silt-sized tephra (USDA Soil Conservation Service 1980). Ash collected immediately after the May 18 eruption will provide an opportunity for future simulation of ashfalls under various environmental regimes (e.g., different seasons, in rain, or on a snow pack). Although not detected in the pollen record (Blinman et al. 1980), simulation may indicate the degree of ecological response produced by relatively low volume ashfalls such as that from Glacier Peak (~11,200–11,300 yr ago). Results of such experimentation, however, will serve only as guides to past vegetation change because no strict

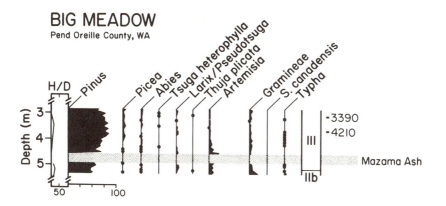

FIG. 11.2   Pollen record in Pollen Zone II at Big Meadow, a fen in Pend Oreille County, Washington, before and immediately after the deposition of Mazama ash (~6700 yr ago). The *Artemisia* pollen taxon increases sharply concomitantly with a temporary decline in *Pinus* pollen. All percentages are based on the nonaquatic sum (Mack et al. 1978).

modern analogs of earlier Holocene communities now occur. In addition, succession and mature community composition in the steppe has been irrevocably changed because of the entry of alien weeds (e.g., *Bromus tectorum*, *Salsola kali*, *Poa pratensis*, and *Cirsium arvense*).

The effects of ash on the steppe in eastern Washington deserve study from another perspective. Subjecting a biological system at any level (population, community, or ecosystem) to external forces permits an examination of the system not possible under homeostatic conditions (Harper 1977). In this context, ash deposition may be one of many possible perturbations of a community that allows determination of its stability. Stability, as used here, is that ability of the community ". . . to weather a stress period or perturbation and return to normal afterward" (Harrison 1979). For example, the steppe of Washington is dominated by caespitose grasses and has shown little stability after disturbance. The creation of arable fields and rangelands and the simultaneous introduction of annual weeds from Eurasia have caused wholesale change in these plant communities (Mack 1981a). Much of this low stability apparently results from the absence of opportunity by these plant species to co-evolve with large numbers of congregating bovids, a source of frequent disturbance. The native caespitose grasses are simply intolerant of frequent heavy grazing and trampling. These plant communities persisted until the advent of agriculture as assemblages of shrubs, perennial forbs, annuals, and caespitose grasses with cryptogams as a soil crust (Mack and Thompson 1982). Destruction of this cryptogam layer which may have exceeded 50% of the total plant cover (Poulton 1955), by large congregating mammals, removed the main defense at the soil surface against the entry of alien weeds. Once established, the aliens have prohibited subsequent succession by native grasses (Daubenmire 1970).

Species replacement may occur in this steppe from changes in surface reflectance and the heat budget because of ash deposition (Stanhill 1965). Instead of mottled dark surfaces with low albedo (dry fallow soils have an albedo of 0.2−0.3), ash coverings raise reflectance. For example, the dry ash at Pullman, Washington, has a surface reflectance of 0.38 (G. S. Campbell, personal communication 1981). Plants with broad horizontal leaves experience warming from below similar to the heating experienced by plants on dry dune sand (Munn 1966). In addition, loss of water from communities may be further exacerbated as water is drawn through silt-sized ash from the soil below and evaporates (W. H. Gardner, personal communication 1980). With low litter production in the *Artemisia*-dominated arid steppe (Mack 1977), these effects may not be quickly ameliorated, and the ash-covered surface can remain intact for a long time.

Cryptogamic crusts, which are destroyed by even minimal livestock treading, can also be destroyed by ash burial. If the mosses, lichens, and algae that form the crust die under the ash, a new episode of succession will begin. While earlier ashfalls may have caused a similar demise of cryptogams, the May 18 eruption produced the first ashfall since the large-scale entry of alien weeds into this region in the late nineteenth century. Consequently, we may see permanent change in community composition even in ungrazed communities through this indirect action of ash.

Although further investigation is awaited, the evidence collected in the 12 months following the May 18, 1980, ashfall has permitted better understanding of the effects of volcanic ash on temperate grassland communities.

MATERIALS AND METHODS

INITIAL RESPONSE OF THE STEPPE PLANTS EMERGENT ON MAY 18

An attempt was made beginning in the week of May 18 to determine the response of plants to the ashfall through the widest possible spec-

trum of steppe vegetation in eastern Washington, using sites south of
the ashfall zone as controls (Mack 1981*b*). Care was taken to distinguish
between ash-induced effects and local variation in phenology among
ash-covered and control sites. Separation of these two categories often
required repeated sampling of equivalent vegetation on ash-covered
and control sites. In spite of extensive synecologic information and a
useful classification for steppe communities in Washington (Daubenmire 1970), the May 18 eruption emphasized that information about
the anthesis, the regularity of annual flowering, and the amplitude of
annual seed maturation of steppe species was often unknown.

MECHANICAL OVERLOADING OF *VERATRUM CALIFORNICUM*

In late May through June, 1981, an attempt was made to isolate those
features that had contributed to the mechanical overloading of *Veratrum californicum* (false hellebore). Above the ground, *Veratrum* consists
of a series of tightly enclosing fleshy leaf sheaths about a small basal
stem. This morphology gives the appearance of a stem, although the
cylindrical leaf sheaths collectively provide little support (Arber 1925).
The height and degree declination from vertical was measured for 14
such "stems" at two sites: one in the *Festuca idahoensis/Symphoricarpos
albus* habitat type (hereafter referred to as h.t.) (SW ¼ Sec. 35, T. 16 N,
R. 43 E, Colfax South quadrangle) and the other in the *Pinus
ponderosa/Symphoricarpos albus* h.t. at the Smoot Hill Biological Reserve
(SE ¼ Sec. 35, T. 16 N, R. 44 E, Albion quadrangle); both sites were in
Whitman County, Washington. Foliar cover was measured by mapping
the perimeter of each plant (as viewed from above) with a mapping
table (Mack and Harper 1977). In this procedure a 2.5 × 2.5 cm wire
mesh cylinder with top and bottom open (1 m tall) enclosed the stem;
the plexiglass mapping table was placed atop the wire mesh cylinder
and was leveled. After the foliar cover was mapped, the wire mesh
cylinder was removed, and dry ash was applied aerially to the plant in
100 g units through a flour sifter to simulate the ashfall. Care was taken
so that the ash was applied over the entire foliar cover, which increased
somewhat as the stem began to tilt. Application of ash continued until
the axis was noticeably tilted and had buckled. Generally the axes were
prostrate within 5 min of the last application of a 100 g unit of ash. In
an earlier experiment, stems toppled in this manner were left in place
for 72 hr before harvesting. No axes rebounded after toppling.
Toppled axes were cut at ground level and the leaves (including leaf
sheaths) were immediately removed and placed in plastic bags until
their return to the laboratory. Axis diameter was measured with ver-

nier calipers, and the stem fresh weight was determined. Leaf sheaths were then dried (80°C, 48 hr) and reweighed after cooling to determine moisture content. The area of leaves flattened under a clear plexiglass plate was measured with a digitizer to the nearest centimeter.

EFFECT OF ASH ON FLOWERING AND FRUITING

In the spring of 1981, the possible effect of ash on pollination and subsequent fruit development was examined for 20 individual flowers of *Iris missouriensis, Geranium viscosissimum, Prunus virginiana, Castilleja cusickii, Rosa woodsii,* and *Geum triflorum* in the *Festuca idahoensis/Symphoricarpos albus* h.t. (NW ¼ Sec. 22, T. 13 N, R. 44 E, Colton quadrangle, Whitman County). On May 27 and 29, ash equivalent to a 2-cm-thick ashfall was applied with a flour sifter to tagged flowers. Flowers at anthesis were selected, and all individuals (both those treated and 20 control flowers) were within a 5 m radius. Flowers were examined within 5 days after ash application; fruits were examined in early August.

DEMOGRAPHY OF SEEDLINGS ON THE ASH

In late September, 1980, all seeds were collected in the micro-depressions within a 5 m radius (except cracks) within a stand in the *Artemisia tridentata/Festuca idahoensis* h.t. (NE ¼ Sec. 15, T. 20 N, R. 35 E, Ritzville NW quadrangle, Adams County) to determine seed composition on the ash surface. The site was stand no. 31 among the unpublished vegetation data collected by R. Daubenmire (1952–1966). In March 1981, documentation of seedling recruitment began in three habitat types at sites along the axis of heaviest deposition of fine-textured ash. The stands were in the *Artemisia tridentata/Agropyron spicatum* h.t. (SW ¼ Sec. 13, T. 19 N, R. 33 E, Schoonover quadrangle, Adams County), the *Artemisia tridentata/Festuca idahoensis* h.t. (location as above), and the *F. idahoensis/Symphoricarpos albus* h.t. (NE ¼ Sec. 34, T. 21 N, R. 40 E, Amber quadrangle, Spokane County). Emergence was monitored routinely by mapping all seedlings in 0.2 × 0.5 m plots using a plexiglass mapping table (Mack and Harper 1977). Control sites (<1 mm of ash deposited) were established in the *Artemisia tridentata/Agropyron spicatum* h.t. on the Arid Lands Ecology (ALE) Reserve at Richland (SE ¼ Sec. 34, T. 11 N, R. 26 E, Iowa Flats quadrangle, Benton County) and in the *F. idahoensis/S. albus* h.t. (NW ¼ Sec. 22, T. 13 N, R. 44 E, Colton quadrangle, Whitman County).

RESULTS AND DISCUSSION

INITIAL RESPONSE OF THE STEPPE
PLANTS EMERGENT ON MAY 18

The timing of the May 18 eruption was fortuitous for observing the maximal effect of a single ashfall on steppe communities in the Pacific Northwest. Most species of vascular plants (including annuals) (Table 11.1) in these communities had resumed vegetative growth, and many were at or near anthesis.

Within 6 weeks following the ashfall, several categories of damage were apparent on those vascular plants emergent on May 18. In general, prostrate or acaulescent species experienced the most conspicuous ash damage through mechanical overloading of stems and leaves. Even with ash removal by wind and rain in June, the distal portions of the leaves of these species continued to be buried under compacted ash throughout the summer. *Balsamorhiza sagittata*, an acaulescent composite with large, triangular-hastate leaf blades, was probably the most widespread example in eastern Washington of ash damage to this morphologic group. Necrosis under the ash resulted in a loss of as much as 49% of the total leaf area of *B. sagittata* in some stands. This loss of photosynthetic area was evident for over half of the usual growing season (early April through August). By comparison, only 27% of the leaf area was lost through herbivory and disease by early August in a control stand. Other low-lying perennial species displaying some leaf death were *Hieracium albertinum*, *Potentilla gracilis*, *Geranium viscosissimum*, and *Geum triflorum*. A much larger group of species displayed ash-induced leaf necrosis in which small flecks of ash remained on leaves through the summer. Damage was most prominent in communities having little surface litter and open canopies. The degree of damage by mechanical overloading and necrosis was directly proportional to the depth of ash received. There was little apparent damage where the uncompacted ash depth was <1 cm (Mack 1981*b*).

Much of the natural vegetation in the Columbia Basin covered by the ashfall is dominated by *Artemisia tridentata* or less often by *Artemisia rigida*. Where the shrub canopy was dense, ash remained on the leaves 1 yr after the eruption. Leaf damage, however, was minimal and ash effects on the shrubs were nil because the large summer leaves of *A. tridentata* usually abscise by late June to early July (Mack 1977). Beyond negligible localized necrosis of leaves, the mature plants of perennial caespitose grasses (e.g., *Agropyron spicatum*, *Festuca idahoensis*, and *Poa sandbergii*) showed no damage from the ashfall, even in areas receiving the heaviest uncompacted deposits (~7 cm). The ash remained at the base of these plants protected from erosion by the densely packed tillers.

The ashfall could have interrupted flowering by destroying pollinators (Cook et al. 1981) and coating the flowers of chasmogamous species. Detrimental effects of the ash on flowers and fruits, however, were probably minimal in most species because flowering continued for several weeks. Flowers were seen on ash-covered plants of *Geranium viscosissimum* near Pullman on May 20, and these plants continued to flower all summer. *Potentilla gracilis* and *Rosa woodsii* also flowered throughout the summer in the same area.

Ash accumulated in both the disc and ray flowers of the erect flowering heads of *Balsamorhiza careyana* and *B. sagittata* and resulted in eventual mechanical overloading. By mid-June, most of the flowering heads in both species had decayed beside the toppled leaves still buried under ash. This burial made unreliable any quantitative comparison of the reduction in seed production in ash-covered stands with that in control populations. Most achenes from plants in the ashfall zone were hollow or decayed; the only *Balsamorhiza* seeds found in the ash-covered area probably developed from the earliest maturing ray and peripheral disc flowers. Results of the simulated ashfall in May 1981 were also consistent with these initial observations. This ashfall had no significant effect on the retention of petals and sepals among the six species examined. Furthermore, by early August there was no significant difference in the level of fruit and seed production between treatment and control plants.

The ash did not interfere with the formation of caryopses in steppe grasses during the summer of 1980. At Ritzville all annual and perennial grasses encountered on the *Artemisia tridentata/Agropyron spicatum* h.t. in July had produced seeds. About 50% of the caryopses of *Festuca octoflora* collected from these sites germinated (observation based on germination in three lots of 100 caryopses each). One gauge of the response of grasses to the ashfall was the generally high production of domestic cereals throughout the ashfall area in the summer of 1980 (Cook et al. 1981).

Along the east flank of the Cascade range, the ashfall consisted primarily of sand-sized particles (Fruchter et al. 1980). This coarse ash did not form a hardened crust and apparently was less hazardous to emergent plants than the fine, silt-sized ash deposited farther east. No leaf necrosis was observed among members of an *Artemisia rigida/Poa sandbergii* community in early August 1980. In addition, no mechanical overloading was observed among plants along the western margin of the steppe.

MECHANICAL OVERLOADING OF
VERATRUM CALIFORNICUM

Veratrum californicum provided one of the most conspicuous examples of mechanical overloading. The broad, plaited leaves of this monocot funneled ash toward the sheathing leaf bases. Although ash poured out from the proximal ends of leaves, the plants 1/3 phyllotaxy ensured that much of the ash was transferred to the next lower leaf. The high water content of the leaf axes (>80% on a dry biomass basis) compounded the plant's susceptibility to toppling (Table 11.2). Height, amount of foliar cover, stem diameter, leaf area and number did not proportionally influence axis toppling, although erect plants (0° declination from vertical) were generally the most resistant to mechanical overloading (Table 11.2). Within the 8–10 hr that ash fell in eastern Washington (Hooper et al. 1980), most individual *V. californicum* plants received the minimum amount of ash to induce permanent axis bending. The additional mass added with rainfall by the end of the first week after the ashfall would have completed the toppling of plants.

As a result of the destruction of the axis, *V. californicum* within the ashfall area was largely prevented from flowering and developing mature capsules. Loss of the 1980 seed crop has probably yielded more serious consequences for false hellebore than for most other species in the steppe because this lily flowers en masse only infrequently (Taylor 1956). The summer of 1980 may have been one of those infrequent years, as evidenced by only two flowering individuals being observed in the study sites during the summer of 1981.

Although the emergent axes of *V. californicum* in the steppe were susceptible to toppling during the ashfall, members of this species were able to emerge through the ash in the Cascade range (Antos and Zobel, Chapter 10, this volume) later in the summer. Such a radically different response emphasizes the importance of phenology in predicting the response of any species (or morphologic group) to an ashfall.

ALTERATION OF THE SOIL SEED BANK

In ashfall areas east of Yakima County, the ash was fine-textured and its compaction by rain has affected seed bank distribution. The seed bank in the soil and seeds cast prior to May 18 (e.g., *Draba verna* and *Lithophragma bulbifera*) were capped by ash. As the ash hardened, the developing crust imposed a new hazard to seedling emergence because species differ considerably in their ability to emerge through a crust (Harper 1977). The seeds of most species, however, were cast after May 18, and the hardened ash surface altered the distribution and potential survival of these disseminules. Seeds previously cast on a surface composed of similarly colored litter with numerous safe sites were much more visible on the uniformly gray background. Furthermore, in the arid steppe, smooth concave microdepressions (<1 cm deep) formed as the ash hardened. Eventually, most seeds cast onto this surface collected in these depressions, creating heterogeneous collections of seeds. Table 11.3 shows the seed composition by late September 1980 in 40 microdepressions in a 5 m radius within a stand on the *Artemisia tridentata/Festuca idahoensis* h.t. Dicot species are poorly represented in this list, although most of the annual dicot species in Table 11.1 were also found at this site. There may have been preferential removal of these seeds by foragers during the summer. In addition, the composition of these seed assemblages undoubtedly was changing through the summer and early autumn as seeds were foraged, lost through blowouts, and augmented by later maturing seeds. Very few seeds remained on the bare hardened ash surface outside these depressions. Although common in all stands in the arid steppe, the seeds of *Bromus tectorum* tended to accumulate more often in cracks than in these microdepressions.

This new distribution of seeds may alter the nature of predation in the arid steppe by granivores, including birds (e.g., the introduced chukar partridge), harvester ants (e.g., *Pogonomyrmex owyheei*; see Willard and Crowell 1965), and small mammals (e.g., *Perognathus parvus, Peromyscus maniculatus, Eutamias minimus, Dipodomys ordii,* and *Reithrodontomys megalotis*; see Kritzman 1977). Both harvester ants (Davidson 1977) and some small rodents (Reichman and Oberstein 1977) selectively forage on the basis of the frequency and density of seeds in the stand. The range of seed dispersions (from individual to clumped groups of seeds) allows a partitioning of the resources among the foraging species. Restriction of seeds to depressions, cracks, or holes in the hardened ash may place the foragers of individual seeds at a selective disadvantage as long as the ash remains unincorporated into the soil-litter interface. A larger fraction of the total seed production on these sites may be removed after the ashfall than before if overall foraging efficiency increases because of increased seed aggregation. Any hypothesis of increased removal, however, assumes that seed distribution prior to ashfall, although presently unknown, was nevertheless not markedly clumped. Data in Table 11.3 indicate that seeds of some species remain in the stand by late September because of failure of foragers to detect all clumps of seeds, satiation of the foragers, or their inability to respond to the new nature of the seed resources. Work is now directed at examining seed removal in ash-covered and control areas.

SEEDLING RECRUITMENT IN HEAVY
ASHFALL AREAS IN THE FIRST YEAR

The pattern of seedling emergence beginning in the autumn of 1980 was largely restricted to small depressions and cracks in the ash surface. Autumn recruitment in the *Artemisia tridentata/Festuca idahoensis* h.t. and the *A. tridentata/Agropyron spicatum* h.t. near Ritzville was not quantified but appeared to be largely confined to seedlings of *Bromus tectorum, Poa* spp., and *Festuca* spp. Very few seedlings germinated on the hardened ash surface outside these depressions (Figs. 11.3 and 11.4).

The following annuals were found in mapped plots in the *A. tridentata/F. idahoensis* h.t. (Adams County) during the spring of 1981: *Draba verna, Microsteris gracilis, Plantago patagonica, Lithophragma bulbifera, Festuca octoflora, F. microstachys, Agrostis interrupta, B. tectorum, Myosurus aristatus,* and *Epilobium paniculatum,* as well as dead dicot and grass seedlings that were too immature to be identified. Censuses in another stand (*A. tridentata/Agropyron spicatum* h.t.) in the thick ash deposition area, also near Ritzville, detected much less diversity in seedlings but did include *Polygonum majus.* These assemblages included winter annuals, such as *B. tectorum,* that had emerged before the first census began in March. On the *F. idahoensis/Symphoricarpus albus* site in Spokane County, seedlings of both perennials and annuals emerged.

Densities of dicot seedlings emergent on the ash in the *A. tridentata/A. spicatum* h.t. and *A. tridentata/F. idahoensis* h.t. were within the range commonly seen in these communities, but only *D. verna* and *M. aristatus* survived to produce seeds. In contrast, all annual grasses (*Agrostis interrupta, B. tectorum, F. microstachys,* and *F. octoflora*) emergent in March 1981 produced abundant seeds in June. In the ash-covered meadow steppe community (*F. idahoensis/S. albus* h.t.), seedlings of *Geranium viscosissimum, Potentilla gracilis, Festuca* spp., and *Poa* spp. emerged but did not survive to July. If this pattern of low dicot seedling survival continues, some species alteration in these communities will occur.

CONSPICUOUS ALGAL GROWTH
IN THE ARID STEPPE

The hardened ash surface in the arid steppe also has supported ephemeral development of a veneer of green and yellow-green algae. Colonization by these aerial immigrants first became noticeable in the late autumn of 1980 on the *Artemisia tridentata/Agropyron spicatum* h.t. and the *A. tridentata/Festuca idahoensis* h.t. near Ritzville; prominence

FIG. 11.3 Seedlings of *Bromus tectorum* emergent from reticulate cracks in the ash (>35 mm thick) 16 km west of Ritzville, Washington, on November 6, 1980. Recruitment was from seeds that had been cast after May 18 and had blown into these developing cracks during the summer, as well as from older seeds on the original soil surface. Very few (<1%) seedlings germinated outside cracks on the hardened ash surface at this site (*Artemisia tridentata/Agropyron spicatum* habitat type).

of algae as a visible surface layer continued into early 1981. The conspicuous growth was apparently restricted to three microsites: under the canopy of *A. tridentata*, along the runways of *Microtus montanus* and surrounding emergent agarics. In each of these microsites, the silt-sized ash held water at a lower matric potential than silt-loam soils of comparable texture. As a result, the rate of drying for ash at Ritzville was consistently slower than that for either quartzite sand or silt-loam soil. This physical property of ash, potentially combined with shading by the *Artemisia* canopy or by frequent additions of water and nitrogen in urine in *Microtus* runways, results in several extra days of log phase growth than would have been possible on silt-loam soil. The maximum algal counts (10^8 cells/g dry soil) occurred in the vole runways, values which rival the highest reported for terrestrial algae. Nevertheless,

FIG. 11.4   Recruitment of grass seedlings (mostly *Festuca* and *Poa* spp. and some *Bromus tectorum*) in swards confined to microdepressions in the hardened ash (~30 mm thick) 11 km north of Ritzville, Washington, on November 6, 1980. Shown are two seedlings that germinated outside a microdepression *(Artemisia tridentata/Festuca idahoensis habitat type)*.

with eventual water loss, the algal vegetative cell viability rapidly declined. An explanation for the algal growth around agarics remains problematic. In addition, the hardened ash surface may have effectively entombed potential algal predators (e.g., nematodes and protozoans), preventing their movement up to the surface veneer of algae. If immigration of these predators into the ash lagged behind that of algae, these "algal blooms" may have partly resulted from predator release (Rayburn et al. 1982).

CRYPTOGAMS

The viability of cryptogams is currently being determined for steppe communities still buried under the ash. Ominous evidence from coniferous forests as far away as the Bitterroot Mountains in northern Idaho reveals that even large mosses are often unable to tolerate entombment under fine-textured ash. Within stands of *Abies grandis*, ash accumulation has been conspicuous within 2 m of grand fir trunks (Mack 1981*b*). Two years after the eruption of Mount St. Helens, *Eurhynchium oreganum*, which is prominent on the forest floor in these stands, has decomposed within this "ash annulus." The moss continues to dominate the cryptogam layer elsewhere in these stands in areas where ash removal from the trees and redeposition on the forest floor has been gradual. Also, there has not yet been much recruitment of new moss gametophytes on the compacted ash. Permanent alteration of plant communities may not be confined to sites of spectacular destruction on or near Mount St. Helens; equally significant vegetation change yet may occur as a result of this cryptogam burial in the steppe.

ACKNOWLEDGMENTS

I thank Rich Old and David A. Pyke who assisted at several stages in collection of the data presented here. Gaylon S. Campbell provided the ash reflectance values. John N. Thompson and Eldon H. Franz provided helpful suggestions and discussions of the manuscript. The research was supported in part by National Science Foundation Grants DEB-8020872 and DEB-8022134 and by the research program of Washington State University.

LITERATURE CITED

ARBER, A. 1925. *Monocotyledons: A Morphological Study*. Cambridge Univ. Press, Cambridge, 258 pp.

BLINMAN, E., P. J. MEHRINGER, and J. C. SHEPPARD. 1980. Pollen influx and the deposition of Mazama and Glacier Peak tephra. In *Volcanic Activity and Human Ecology*, P. D. Sheets, and D. K. Grayson (eds.), Academic Press, New York, pp. 393–425.

COOK, R. J., J. C. BARRON, R. I. PAPENDICK, and G. J. WILLIAMS. 1981. Impact on agriculture of the Mount St. Helens eruptions. *Science* 211:16–22.

DAUBENMIRE, R. 1952–1966. Vegetation of eastern Washington and northern Idaho: Field records of vegetation and soils. I. Steppe stands 1–100. In *Manuscripts, Archives, Special Collections*. Washington State Univ. Libraries, Pullman.

———. 1970. Steppe vegetation of Washington. *Wash. State Agr. Exp. Sta. Tech. Bull. 62*, 131 pp.

———. 1978. *Plant Geography*. Academic Press, New York, 338 pp.

DAVIDSON, D. W. 1977. Foraging ecology and community organization in desert seed-eating ants. *Ecology* 58:725–737.

FRUCHTER, J. S., D. E. ROBERTSON, J. C. EVANS, K. B. OLSEN, E. A. LEPEL, J. C. LAUL, K. H. ABEL, R. W. SANDERS, P. O. JACKSON, N. S. WOGMAN, R. W. PERKINS, H. H. VAN TUYL, R. H. BEAUCHAMP, J. W. SHADE, J. L. DANIEL, R. L. ERIKSON, G. A. SEHMEL, R. N. LEE, A. V. ROBINSON, O. R. MOSS, J. K. BRYANT, and W. C. CANNON.

1980. Mount St. Helens ash from the 18 May 1980 eruption: Chemical, physical, mineralogical, and biological properties. *Science* 209:1116–1125.

GODLEY, E. J. 1975. Flora and vegetation. In *Biogeography and Ecology in New Zealand*, G. Kuschel (ed.), Dr. W. Junk, The Hague, pp. 117–229.

GRIGGS, R. F. 1918. Scientific results of the Katmai expeditions of the National Geographic Society. I. The recovery of vegetation at Kodiak. *Ohio J. Sci.* 19:1–57.

HARPER, J. L. 1977. *The Population Biology of Plants*. Academic Press, New York, 892 pp.

HARRISON, G. W. 1979. Stability under environmental stress: Resistance, resilience, persistence and variability. *Amer. Natl.* 113:659–669.

HITCHCOCK, C. L., A. CRONQUIST, M. OWNBEY, and J. W. THOMPSON. 1955, 1959, 1961, 1964, 1969. *Vascular Plants of the Pacific Northwest*. Parts 1–5. Univ. of Washington Press, Seattle.

HOOPER, P. R., I. W. HERRICK, E. R. LASKOWSKI, and C. R. KNOWLES. 1980. Composition of the Mount St. Helens ashfall in the Moscow–Pullman area on 18 May 1980. *Science* 209:1125–1126.

KRITZMAN, E. B. 1977. *Little Mammals of the Pacific Northwest*. Pacific Search Press, Seattle. 120 pp.

MACDONALD, G. A. 1972. *Volcanoes*. Prentice-Hall, Englewood Cliffs, New Jersey. 510 pp.

MACK, R. N. 1977. Mineral return via the litter of *Artemisia tridentata*. *Am. Midl. Natur.* 97:189–197.

———. 1981a. Invasion of *Bromus tectorum* L. into western North America: An ecological chronicle. *Agro-Ecosys.* 7:145–165.

———. 1981b. Initial effects of ashfall from Mount St. Helens on vegetation in eastern Washington and adjacent Idaho. *Science* 213:537–539.

MACK, R. N., and J. L. HARPER. 1977. Interference in dune annuals: Spatial pattern and neighbourhood effects. *J. Ecol.* 65:345–363.

MACK, R. N., N. W. RUTTER, V. M. BRYANT, and S. VALASTRO. 1978. Late Quaternary pollen record from Big Meadow, Pend Oreille County, Washington. *Ecology* 59:956–966.

MACK, R. N., and J. N. THOMPSON. 1982. Evolution in steppe with few large, hooved mammals. *Am. Naturalist* 119:757–773.

MUNN, R. E. 1966. *Descriptive Micrometeorology*. Academic Press, New York. 245 pp.

POULTON, C. E. 1955. Ecology of the non-forested vegetation in Umatilla and Morrow Counties, Oregon. Ph.D Thesis, Washington State Univ., Pullman.

QUINN, R., and M. FOLSOM. 1980. The downwind distribution of Mount St. Helens ash, May 18th eruption. In *Proceedings of Washington State University's Conference on the Aftermath of Mount St. Helens*, J. J. Cassidy (chairman), July 8–9, 1980, Pullman, Washington, pp. 51–57.

RAYBURN, W. R., R. N. MACK, and B. METTING. 1982. Conspicuous algal colonization of the ash from Mount St. Helens. *J. Phycology* 18(4):537–543.

REICHMAN, O. J., and D. OBERSTEIN. 1977. Selection of seed distribution types by *Dipodomys merriami* and *Perognathus amplus*. *Ecology* 58:636–643.

SMATHERS, G. A., and D. MUELLER-DOMBOIS. 1974. Invasion and recovery of vegetation after a volcanic eruption in Hawaii. *National Park Service Scient. Mono. Series No. 5*, 129 pp.

STANHILL G. 1965. Observations on the reduction of soil temperature. *Agr. Meteorol.* 2:197–203.

TAYLOR, C. A. 1956. The culture of false hellebore. *Econ. Bot.* 10:155–165.

USDA SOIL CONSERVATION SERVICE. 1980. *Mount St. Helens Ash Fallout Impact Assessment Report*. Spokane, Washington, 78 pp.

WILLARD, T. R., and H. H. CROWELL. 1965. Biological activities of the harvester ant, *Pogonomyrmex owyheei*, in central Oregon. *J. Econ. Entomol.* 58:484–489.

TABLE 11.1

FLOWERING PHENOLOGY OF ANNUAL FORBS AND GRASSES IN THE
Artemisia tridentata/Agropyron spicatum ZONE AT
RITZVILLE, WASHINGTON[a]

Species	Some seed cast by May 18	Most seed cast by May 18
Grasses		
Agrostis interrupta L.	X	
Bromus tectorum L.		
Deschampsia danthonioides (Trin.) Munro		
Festuca microstachys Nutt.	X	
Festuca octoflora Walt.	X	
Forbs		
Amsinckia lycopsoides Lehm.		
Athysanus pusillus (Hook.) Greene	X	
Camelina microcarpa Andrz.		
Chorispora tenella (Pall.) DC.	X	
Collinsia parviflora Lindl.	X	
Descurania richardsonii (Sweet) Schulz	X	
Draba verna L.		X
Erysimum repandum L.	X	
Holosteum umbellatum L.		X
Lithophragma bulbifera Rydb.		X
Lithophragma parviflora (Hook.) Nutt.	X	
Microsteris gracilis (Hook.) Greene	X	
Myosurus aristatus Benth.	X	
Plantago patagonica Jacq.		
Plectritis macrocera T. & G.	X	
Sisymbrium altissimum L.		
Stellaria nitens Nutt.	X	

[a]By May 18, 1980, all these species had begun flowering and some seed dispersal had begun in ~50% of the group. *Draba verna*, *Holosteum umbellatum*, and *Lithophragma bulbifera* cast most seed by mid-May. Breadth of the flowering season was compiled from Hitchcock et al. (1955, 1959, 1961, 1964, 1969) and from herbarium specimen data.

TABLE 11.2
CHARACTERISTICS OF THE INDIVIDUAL "STEMS" OF *Veratrum californicum* SUBJECTED TO A
SIMULATED ASHFALL NEAR ALBION, WASHINGTON, MAY–JUNE 1981

Stem number	Stem height (cm)	Foliar cover (cm)	Leaf area (cm)	Stem diameter (mm)	Leaf number	Stem angle (°)[b]	Stem moisture (%)	Ash to topple (g)[b]
1	60	244	1158	14.0	14	0	84.6	252
2	63	448	1735	15.5	12	14	84.7	200
3	68	390	1629	12.7	11	0	84.0	300
4	71	418	1555	16.0	12	10	85.5	200
5	75	583	3080	24.0	13	0	85.2	600
6	79	726	3271	25.0	15	14	86.0	300
7	80	413	1682	18.3	12	0	85.5	900
8	87	511	2152	19.1	11	14	87.0	157
9	92	583	3410	28.3	15	0	86.5	900
10	96	1006	4319	23.0	15	18	87.2	300
11	97	973	2961	26.5	14	29	88.5	200
12	100	703	2778	24.0	11	10	87.8	300
13	102	835	4092	27.7	13	10	87.0	600
14	102	624	6287	36.1	20	0	86.5	2500

[a]Declination from vertical.
[b]All stems were permanently toppled by the amount of ash shown.

TABLE 11.3

Seed Composition in Late September 1980 of 40 Microdepressions in a 5 m Radius in the *Artemisia tridentata*/*Festuca idahoensis* Habitat Type near Ritzville, Washington

Plot Number	*Bromus tectorum*	*Agrostis interrupta*	*Festuca* spp.	*Lomatium* spp.	*Plagiobothrys tenellus*
1	12	—	644	—	—
2	6	1	168	—	—
3	—	1	234	—	—
4	—	6	84	—	—
5	—	7	34	—	—
6	1	23	39	—	—
7	1	2	91	—	—
8	8	5	185	—	—
9	1	15	138	—	—
10	—	11	106	—	—
11	—	1	161	—	—
12	—	3	136	—	—
13	1	47	193	—	—
14	—	86	281	—	—
15	1	—	186	1	—
16	—	2	191	—	—

17	—	—	135	—
18	—	—	135	147
19	—	1	57	7
20	—	—	65	—
21	—	—	178	19
22	—	1	183	15
23	—	—	98	—
24	—	—	127	3
25	—	—	88	—
26	—	—	125	56
27	3	2	74	—
28	—	1	356	—
29	—	—	275	2
30	—	—	14	1
31	—	6	104	4
32	—	—	79	2
33	—	1	55	14
34	—	1	87	6
35	—	—	196	6
36	—	—	48	—
37	—	—	58	8
38	—	—	112	—
39	—	—	305	2 · 5
40	—	—	166	167

12

Persistence of Mount St. Helens Ash on Conifer Foliage*

D. E. Bilderback and J. H. Slone

ABSTRACT

During the eruption of Mount St. Helens on May 18, 1980, varying amounts of ash from the volcanic plume were deposited on the coniferous forest ecosystems of Washington, Idaho, and Montana. Almost 150 days later, needles of Douglas fir (*Pseudotsuga menziesii*) and Engelmann spruce (*Picea engelmanni*), growing at sites that had received 1.0 cm or more of ash, were still heavily coated with 382–543 mg of ash per gram needle dry weight. Little variation in the amount of persistent ash on the needles of Douglas fir occurred over a distance of 220 km from the volcano. In contrast, little ash persisted on the foliage of ponderosa pine (*Pinus ponderosa*) and white bark pine (*Pinus monticola*) growing in the same localities as Douglas fir and Engelmann spruce. The needles of Douglas fir growing in areas that received <1.0 cm of ash also retained little or no ash. The surfaces of ash-laden needles of Douglas fir had significantly more epiphytic fungal colonies and hyphae associated with them than did needles with little or no ash. Needles of Engelmann spruce coated with heavy deposits of ash also had significantly more fungal colonies than did needles with less ash. The ash-laden needles created a modified environment apparently more favorable for the growth of epiphytic fungi. The adherence of the ash particles to the mucilage produced by the fungal colonies and hyphae stabilized the ash layer so that it became a persistent feature on the surface of Douglas fir and Engelmann spruce needles. The ash had no apparent effect on the internal anatomy of Douglas fir needles. Eight weeks after being artificially dusted with ash, needles had significantly less nonstructural carbohydrates than did needles from the control plants. Ash-coated and control seedlings had similar dark respiration rates, compensation points, and light saturation levels. Seedlings artificially coated with ash, however, exhibited an enhanced photosynthetic rate.

*This work was supported by a U.S. Forest Service Cooperative Agreement INT-80-106-CA.

INTRODUCTION

The explosive eruption of Mount St. Helens on May 18, 1980, propelled a large ash plume into the atmosphere to an altitude of 20−25 km (Rosenfeld 1980). After 8 hr, the plume had reached an estimated volume of 2×10^6 km^3 (Danielsen 1981) and covered most of eastern Washington, northern Idaho, and western Montana. An estimated 1.5−2.0 km^3 ash fell on farm, range, and forest lands (Hammond 1980). The impact of ash on the agriculture of eastern Washington and northern Idaho has been adequately described by Cook et al. (1981). Except for a brief discussion in an article by del Moral (1981), the impact of volcanic ash on conifers has not been investigated.

MATERIALS AND METHODS

Branchlets were collected from various genera and species of conifers growing at 22 sites in Washington, Idaho, and Montana on October 10 and 11, 1980. The amount of deposited ash and the location of the sampling sites are shown in Figure 12.1. The distance of the sites from Mount St. Helens, the elevation, and the vegetation type at the sites are given in Table 12.1. Sites 1−7 were located in the Snoqualmie and

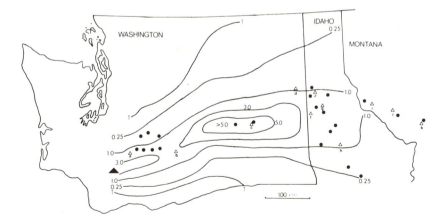

FIG. 12.1   Depth of volcanic ash (cm), location of sampling sites (●) (see Table 12.1), and location of selected climatological sites (△) (see Table 12.2). Map adapted from Cook et al. (1981).

Gifford Pinchot National Forests at a distance of 70–119 km from the volcano. Five of the sites (1, 2, 3, 5, and 7) were located in areas that received moderate to heavy deposits of ash. Two sites (4 and 6) were exposed only to a light dusting of ash. All seven sites had Douglas fir (*Pseudotsuga menziesii*), and ponderosa pine (*Pinus ponderosa*) grew at Sites 3, 5, 6, and 7. Located at a higher elevation, Site 2 had western hemlock (*Tsuga heterophylla*), lodgepole pine (*Pinus contorta*), and Cascade fir (*Abies amabilis*). Since the distance from the volcano did not vary more than 32 km nor the site elevation more than 354 m, Sites 1 and 2 were compared to Site 4 and Sites 3, 5, and 7 were compared to Site 6.

Sites 8 and 9 were located in a region of central Washington that received >5.0 cm of uncompacted ash. This area is sagebrush steppe and is thus too arid to support stands of conifers. A few branchlets of Douglas fir, ponderosa pine, and Scots pine (*Pinus sylvestris*), however, were collected from trees planted near an interstate highway rest stop and along a public right-of-way.

Ten sample sites were located in the panhandle of Idaho. Five of these sites were in or adjacent to the Clearwater, St. Joe, or Coeur d'Alene National Forests. Both Douglas fir and ponderosa pine grew at Sites 10, 11, and 12 (Table 12.1), which were similar to one another. Sites 10 and 11, however, received a moderate to heavy deposit of ash, whereas Site 12 was subjected to a much lighter dusting. Ponderosa pine and Douglas fir grew at Sites 13, 15, 18, and 19. Sites 13 and 15 received moderate amounts of ash, and Sites 18 and 19 were exposed only to a light dusting. Engelmann spruce (*Picea engelmanni*), western red cedar (*Thuja plicata*), grand fir (*Abies grandis*), subalpine fir (*Abies lasiocarpa*), white pine (*Pinus monticola*), lodgepole pine, and ponderosa pine grew at Sites 14, 16, and 17. All these sites received heavy deposits of ash. Of the three sites in Montana, Site 20 was at the highest elevation, that being 1478 m. Engelmann spruce and lodgepole pine at this site were exposed to moderate amounts of ash. Ponderosa pine and Douglas fir at Sites 21 and 22 received a relatively light dusting of ash.

The locations of climatological stations are shown in Figure 12.1. Precipitation recorded at those stations from May 18 to October 10, 1980, is summarized in Table 12.2.

At all sites, branchlets were removed from trees 1.9–2.5 m above ground level. All sampled branchlets had been fully exposed to the ash fallout occurring on May 18, 1980. To reduce any possible roadside disturbance, trees were sampled at some distance from the highway. The branchlets were placed directly into plastic bags with appropriate labels. On returning to the University of Montana, the samples were stored at 4°C in a refrigerator.

For a qualitative determination of the ash retained on the surface of needles, representative branchlets from the various sites were photographed on a copy stand. The central portion of the abaxial and adaxial surfaces of representative Douglas fir needles were also photographed at a magnification of ×15, using a dissecting microscope.

For scanning electron microscopic (SEM) studies, a 1.0-cm-long central portion was removed from representative needles of Douglas fir, ponderosa pine, lodgepole pine, and Engelmann spruce from various sites. Care was taken not to dislodge the ash deposits from the surface of the needles. The samples were first air dried and then dried at 50°C for 1 hr before being affixed to studs and coated for examination with a Zeiss Novascan 30 SEM. The more standard techniques of fixation and dehydration were not used because they would have dislodged much of the ash. Both the abaxial and adaxial surfaces of the needles were examined with the SEM. The ash was intentionally removed from the surfaces of some fresh needles by washing them gently with water and a small paintbrush. These needles were then oven dried at 90°C for 12 hr before being examined with the SEM.

To determine the amount of persistent ash on the needles, 20 representative needles of Douglas fir or Englemann spruce from each site were washed gently with water and a small paintbrush. For dry weight determinations, the needles were oven dried at 90°C for 12 hr. The ash washed from the needles was placed, using suction, onto preweighed Millipore filters. The ash-laden filters were oven dried and weighed for the determination of the amount of persistent ash per gram of needle dry weight.

To investigate the extent of epiphytic fungi on the surface of needles, a 1.0-cm section was removed from the central portion of at least nine randomly selected 2-yr-old needles of Douglas fir and Engelmann spruce. When the adaxial surfaces of the sections were moistened, black fungal colonies became visible with a dissecting microscope at ×8 magnification. When the colonies were too numerous to count with the dissecting microscope, the needle sections were photographed at a ×5 magnification and number of colonies were determined from 35-mm negatives projected onto a piece of white paper.

To identify the dominant genera of epiphytic fungi, Douglas fir needles from Sites 10 and 11 were rinsed 10 times with sterile distilled water. Using sterile techniques, fungal colonies and pieces of hyphae were transferred from the washed needle surface to the surface of a 2% malt agar medium in small petri dishes. The fungi were incubated at 25°C in continuous darkness. The fungi were periodically subcultured, and small portions of the cultures were removed for examination with the light microscope or were air dried for study with the SEM.

To determine the anatomical or morphogenetic effects of volcanic ash on Douglas fir needles, portions of needles were fixed in 3% gluteraldehyde buffered with 0.05 M potassium phosphate of pH 6.8. The tissues were dehydrated in a graded ethanol series, embedded in the medium of Feder and O'Brien (1968), and sectioned at $1-3 \mu$m. Sections were stained with periodic acid–Schiff for carbohydrates (Feder and O'Brien 1968) and with aniline blue-black for proteins (Fisher 1968).

To determine the effects of an artificial dusting of volcanic ash on carbohydrate accumulation, young seedlings of Douglas fir were obtained on two different occasions from the U.S. Forest Service. The first group of seedlings ranged in height from 7 to 12 cm, whereas the height of the second group was $18-19$ cm. Randomly selected seedlings were placed at random in a chamber and coated with volcanic ash according to the modified methods of Atkins et al. (1954). Randomly selected control seedlings, not coated with ash, were also subjected to the same conditions in the chamber as were the ash-coated plants. The amount of ash deposited on the foliage averaged 397 mg per gram needle dry weight, and the ash used to artificially coat the seedlings was collected in Missoula on May 18, 1980. The first group of seedlings was transferred to and maintained in a greenhouse for 3 weeks. So as not to dislodge the ash from the foliage, plants were watered once per week with 30 ml from a plastic squeeze bottle. The second group of seedlings was placed in a growth chamber with 16 hr of light per day at an intensity of 30μ Einstein/m^2/sec^1 for 8 weeks and watered in a similar manner.

To determine the amount of nonstructural carbohydrates of needles, six plants each of ash-treated and control seedlings were clipped at midafternoon. The ash was washed from needles by agitating the stems and needles for 1 min in 20 ml of distilled water. Needles were stripped from the stems, dried at 90°C for 12 hr, and ground to pass through a 60-mesh screen. The nonstructural carbohydrates were quantified using the method of daSilveira et al. (1978). Because no large differences in carbohydrates were observed when ash-coated plants were compared to controls and because ash-coated needles were difficult to section, carbohydrates were not qualitatively determined.

To investigate photosynthesis by artificially ashed and control seedlings, an open system gas exchange apparatus was utilized. Three seedlings of Douglas fir were sealed in a ventilated, temperature-controlled, teflon-coated aluminum assimilation chamber with a glass top. A constant flow of air from a compressed air cylinder was passed through the chamber to a Beckman 215 infrared gas analyzer. Flow rates were maintained so that the CO_2 concentration in the chamber did not drop below 300μ l. The difference between the CO_2 concentration of the compressed air and that of the air stream leaving the chamber was taken as the amount of CO_2 fixed by photosynthesis. Light was supplied by a General Electric quartz lamp, and the photosynthetically active radiation was measured with a LiCor quantum meter. To measure light-dependent photosynthesis, irradiance was decreased in increments by covering the glass lid of the chamber with neutral density screens. The air temperature of the chamber was measured with a thermocouple and was held at 25°C for the various light intensities. Photosynthetic rates were expressed on a needle gram dry weight basis and represented an average of replicate samples of groups of three plants.

RESULTS

SITES NEAR MOUNT ST. HELENS

Douglas fir needles from sites located between the 1.0 and 3.0 cm ash isograms retained from 84 to 430 mg of ash per gram of needle dry weight (Figs. 12.2 and 12.3A). The ash coated both the abaxial and adaxial surfaces of the needles (Fig. 12.3B and C). All age classes of Douglas fir needles retained ash except at Site 1 where the current year's needles had less ash than did older needles on the same branch-let. The ash also coated the older needles of the Cascade fir and western hemlock growing at Site 2. Usually, however, ponderosa pine or lodgepole pine needles from the four sites retained little or no ash (Fig. 12.3D and E). Needles from Douglas fir growing at Site 7 situated between the 0.25- and 1.0-cm ash isogram had only a small amount of ash (20 mg/g needle dry weight). No ash was observed on the foliage of ponderosa pine at this site. When sites were subjected to 0.25 cm or less ash, little or no ash persisted on the needles of Douglas fir.

The needles of Douglas fir growing at Site 1 had significantly more black epiphytic fungal colonies than did needles from Site 4 (Fig. 12.4). There was no significant difference in the number of fungal colonies on Douglas fir needles from Sites 3 and 6 or from Sites 5 and 7. The needles with persistent ash, however, always had a larger number of hyaline hyphae growing among the ash or tephra (Fig. 12.3F). These needles also had significantly more black hyphae than did needles without persistent ash.

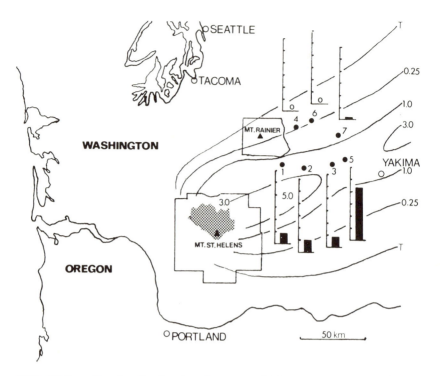

FIG. 12.2   Depth of volcanic ash (cm) and histograms representing the amount of ash retained by Douglas fir needles. Each unit on the histograms represents 100 mg of ash per gram of needle dry weight. Map adapted from Findley (1981).

CENTRAL WASHINGTON SITES

The two central Washington sites (8 and 9) received deposits of ash >5.0 cm in depth. The age of the needles on the Scots pine made a significant difference in the amount of retained ash. The newer current year's needles had little ash, whereas the 2-yr-old foliage was heavily coated with ash. The Douglas fir needles retained 382 mg of ash per gram needle dry weight. Both surfaces of needles of all ages were encased in ash. Beneath the ash, the adaxial surface of the Douglas fir needles had only seven black fungal colonies. The foliage of ponderosa pine retained no visible ash.

IDAHO SITES

The foliage of Douglas fir growing at Sites 10 and 11 retained an average of 422 mg of ash per gram needle dry weight (Fig. 12.5). Both surfaces of needles of all ages were heavily encrusted with ash (Fig. 12.3G and H). In contrast, foliage from Douglas fir trees growing at Site 12 on the other side of the 1.0-cm ash isogram had only 15 mg of ash per gram needle weight (Fig. 12.3I and J). The needles of ponderosa pine growing at these three sites had no persistent ash.

The number of epiphytic fungal colonies growing beneath the ash on the surface of Douglas fir needles from Sites 10 and 11 was significantly greater than those on foliage from Site 12 (Fig. 12.6). At Site 12 the fungal colonies were restricted to the adaxial groove of the needle, but at Sites 10 and 11 the colonies were more evenly distributed on the adaxial surface of the heavily ash-laden needles. On the abaxial surface of the Douglas fir needles, conidia of *Cladosporium cladosporioides* could occasionally be seen among the ash particles (Fig. 12.3K). Removal of the ash from the needles revealed large numbers of ash-coated fungal colonies and hyphae on the adaxial needle surface (Fig. 12.3L). When the ash was removed from the abaxial surface, a well-developed mycelial mat with multicellular conidia could be seen above the stomata (Fig. 12.3M and N). At higher magnification, fibrils could be seen traversing the space between the hyphae and adjacent ash particles (Fig. 12.3O). Ponderosa pine needles had no fungal colonies and only a small amount of hyphae.

Sites 13–17 were located at higher elevations (842–895 m) than were Sites 10–12. Douglas fir needles from Sites 13 and 15 retained only a moderate amount of ash (44–54 mg/g needle dry weight) but did have a large number of epiphytic fungal colonies growing on their adaxial surfaces (Figs. 12.5 and 12.6). The needles of ponderosa pine from the same sites had neither ash nor a significant number of epiphytic fungi.

Engelmann spruce, western red cedar, grand fir, subalpine fir, white bark pine, and lodgepole pine grew at Sites 14, 16, and 17. Regardless of age, all of the needles of Engelmann spruce were heavily coated with ash (94–540 mg/g needle dry weight) (Fig. 12.5). The older foliage of the firs and western red cedar also had some persistent ash deposits. There was no visible ash on the needles of the white bark pine. Some of the older needles on the branchlets of lodgepole pine had ash; the current year's foliage lacked ash deposits.

All four surfaces of ash-laden Engelmann spruce needles had a large number of epiphytic fungal colonies (Fig. 12.6 and 12.3P). Needles from Sites 14 and 16 had more ash and significantly more colonies than did needles from Site 17. The older needles of grand fir and lodgepole pine also had numerous fungal colonies.

The foliage from Douglas fir growing at Sites 18 and 19 retained no ash and had few epiphytic colonies (Figs. 12.5 and 12.6).

MONTANA SITES

Site 20 was a high elevation Engelmann spruce/lodgepole pine site. The needles of the spruce retained 77 mg of ash per gram needle dry weight and had a large number of fungal colonies (Figs. 12.5 and 12.6). The foliage of the lodgepole pine was also heavily coated with ash and had numerous fungal colonies.

Sites 21 and 22 received 0.15–1.0 cm of ash. The Douglas fir foliage had no persistent ash and only a few epiphytic fungal colonies in the adaxial groove of the needle.

FUNGAL IDENTIFICATION

The dominant genera of epiphytic fungi growing on the adaxial surface of the needles of Douglas fir at Sites 10 and 11 were *Hormonema dematioides* and an unidentified Loculoascomycete with numerous blastoconidia. On the abaxial needle surface, the dominant genera were *Monodictys* and *Leptographium*.

FIG. 12.3   A. A portion of a Douglas fir branchlet from sample Site 5 showing ash deposits on the needles, ×1. B. Adaxial surface of a needle of Douglas fir from Site 5 showing ash deposition, ×25. C. An ash-laden abaxial surface of a needle of Douglas fir from Site 5, ×25. D. Adaxial surface of a needle of ponderosa pine from Site 5. A small amount of ash is visible between the stomatal bands, ×25. E. Abaxial surface of needle of ponderosa pine from Site 5, ×25. F. SEM view of fungal hyphae among the ash particles on a Douglas fir needle from Site 5, ×127. G. Adaxial surface of an ash-laden needle of Douglas fir from Site 10, ×25. H. Abaxial surface of a needle of Douglas fir from Site 10, ×25. I. Adaxial surface of a Douglas fir needle from Site 12, ×25. J. Abaxial surface of a needle of Douglas fir from Site 12, ×25. K. SEM view of *Cladosporium* conidia and hyphae in the ash layer on the abaxial surface of a Douglas fir needle from Site 11, ×950. L. SEM view of the adaxial surface of a Douglas fir needle from Site 10 with the ash removed showing numerous fungal colonies and hyphae, ×75. M. SEM view of the abaxial surface of a Douglas fir needle from Site 10 with the ash removed showing the hyphal mat over the stomata (arrow), ×43. N. SEM view of conidia associated with the hyphae on the abaxial surface of a Douglas fir needle from Site 10, ×790. O. SEM view of a fungal hypha attached to ash particles by fine fibrils (arrows); abaxial surface of a Douglas fir needle from Site 10, ×3500. P. SEM view of ash-coated fungal colonies on the surface of an Engelmann spruce needle from Site 14; the ash has been removed, ×560.

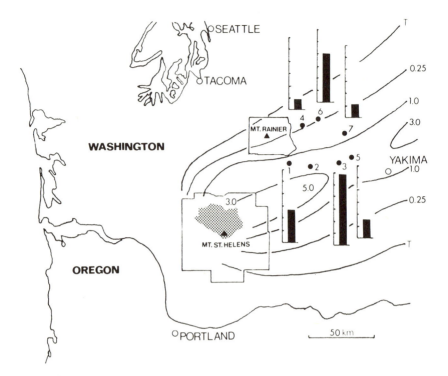

FIG. 12.4   Depth of volcanic ash (cm) and histograms representing the
number of black epiphytic fungal colonies growing on the adaxial surface of
Douglas fir needles. Each unit on the histograms represents 10 colonies per
centimeter of needle. Map adapted from Findley (1981).

ANATOMICAL AND PHYSIOLOGICAL
EFFECTS

The volcanic ash had no anatomical or morphogenetic effects on
Douglas fir needles. The internal anatomy of needles coated with ash
did not differ from that of control needles.

After 3 weeks in a greenhouse, the amount of nonstructural
carbohydrates of needles from artificially ash-coated seedlings did not
differ significantly from the carbohydrates of needles from control
plants (Table 12.3). There was a trend, however, toward a decrease in
nonstructural carbohydrate in the ash-coated needles. When seedlings
were maintained in a growth chamber for 8 weeks, needles artificially
coated with ash exhibited a significant decrease in the amount of
nonstructural carbohydrate when compared to the control.

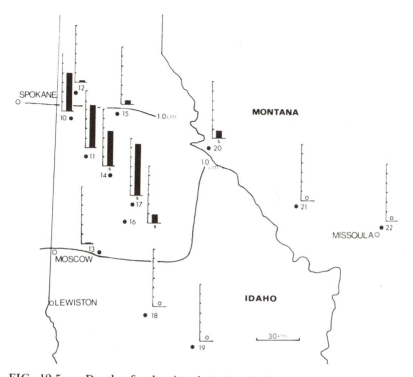

FIG. 12.5   Depth of volcanic ash (cm) and histograms representing the amount of ash retained by Douglas fir and Engelmann spruce needles (s). Each unit on the histograms represents 100 mg of ash per gram of needle dry weight. Ash depth taken from Cook et al. (1981).

When photosynthesis of artificially ashed seedlings was compared to that of control plants, it was found that the two groups of seedlings exhibited similar dark respiration rates, compensation points, and light saturation levels (Fig. 12.7). Seedlings coated with volcanic ash, however, did fix CO_2 at a faster rate than did the control seedlings.

DISCUSSION

On May 18, 1980, large amounts of ash from the explosive eruption of Mount St. Helens descended on the diverse forest ecosystems of Washington, Idaho, and Montana. Almost 150 days later, the needles of Douglas fir and Engelmann spruce growing at locations that received

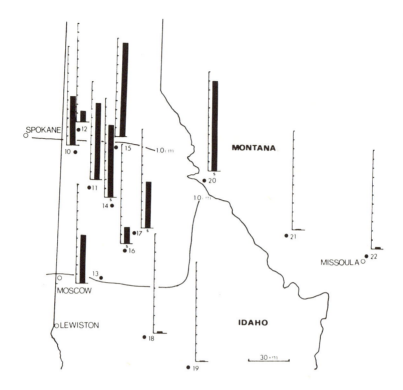

FIG. 12.6   Depth of volcanic ash (cm) and histograms representing the number of black epiphytic fungal colonies growing on the adaxial surface of Douglas fir needles and on one of the four surfaces of Engelmann spruce needles (s). Each unit on the histograms represents 100 colonies per centimeter of needle. Ash depth taken from Cook et al. (1981).

at least 1.0 cm of ash were still heavily coated with volcanic ash. To a lesser extent, western red cedar, Cascade fir, grand fir, and subalpine fir needles also had persistent ash deposits. Although the physical characteristics of the tephra did vary with distance from the volcano (Fruchter et al. 1980), the amount of ash retained by Douglas fir foliage did not. Needles from trees growing as close as 113 km and as far as 434 km from the volcano were coated with as much as 430–440 mg of ash per gram of needle dry weight. Foliage from sites receiving 1.0–3.0 cm of ash actually retained more ash than did needles of trees growing at a location receiving >5.0 cm of tephra. The greatest amount of ash retained by any conifer foliage (540 mg/g needle dry weight) was observed on the needles of Engelmann spruce growing at a site 456 km from Mount St. Helens.

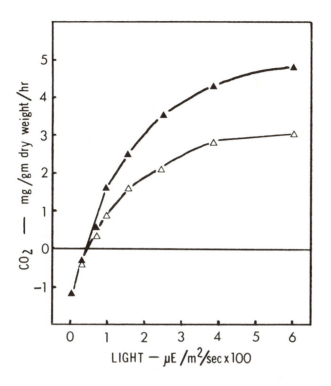

FIG. 12.7   Photosynthesis of seedlings 8 weeks after being artificially coated with volcanic ash (▲) and control seedlings (△). E = Einstein.

In stark contrast to the ash-laden foliage of Douglas fir and Engelmann spruce, the needles of ponderosa pine and white bark pine growing at the same locations retained little or no ash. As might be expected, the needles of Douglas fir growing at sites that received <1.0 cm of ash also had little or no ash.

The persistence of ash on the foliage of Douglas fir and Engelmann spruce could not be explained by the climatology of the region (Table 12.2). All locations with ash-coated needles received a total of 6.86–28.37 cm of precipitation during the 150-day period after the eruption of the volcano. This amount of precipitation should have dislodged the ash from the surface of the needles. Closer examination of the climatological data, however, revealed that all sites except one received only gentle rainfall, which presumably could not effectively dislodge the ash. In fact, the gentle rainfall may have caused some redistribution of ash from the adaxial to the abaxial surface of the needle.

The retention of ash by Douglas fir and not by the neighboring ponderosa pine was most intriguing. The more horizontal needles of the Douglas fir could support more ash deposition than the vertical needles of the ponderosa. Some of the ash may have initially adhered to the gelatinous epiphytic fungi growing in the midrib groove and near the stomata of the Douglas fir needle. Since the fungi occupy only a small area of the needle surface (Carroll 1979), it was highly unlikely that there were enough mucilaginous hyphae on the needle surfaces to account for the initial retention of large amounts of ash. In fact, needles of trees growing in central Washington retained large amounts of ash but had only a few epiphytic colonies. In all likelihood, the ash initially interacted with and adhered to the cuticle of Douglas fir needles in a different manner than it did with the needles of ponderosa pine. When young epiphyte-free seedlings of Douglas fir were artificially dusted with volcanic ash, not all of the ash could be dislodged by simply washing the needles with running water (S. Marvel, personal communication 1981).

Needles of Scots pine and the true firs exhibited an age-dependent retention of ash. The younger, current year's needles had significantly less ash than did the older needles. The younger needles were most likely still enclosed in the buds or had just begun to expand when the ash fell. The smaller amounts of ash on the surface of the younger needles presumably resulted from secondary deposition by wind.

Even though the ash initially clung to the surfaces of crop plants, most of the ash was eventually shed from the leaf surfaces (Cook et al. 1981). With conifers, del Moral (1981) reported that western red cedar growing just a few kilometers from the blast zone retained ash more tenaciously than did Douglas fir. His observations seemingly contradict those made during this investigation. Perhaps the physical nature of the ash near the volcano was very different from that deposited farther away.

The ash-laden needles of Douglas fir growing on the eastern slope of the Cascade range had significantly more black hyphae on their surfaces than did needles without persistent ash. These hyphae, as well as hyaline hyphae, penetrated the ash layer and were frequently observed growing among the ash particles.

A large number of black epiphytic colonies of *Hormonema* and other fungi grew under the ash on the adaxial surface of Douglas fir needles from the Idaho sites. Needles with little or no ash had significantly fewer epiphytic colonies. Epiphytic fungal colonies on Engelmann spruce needles could be found on all four surfaces of the needles, and needles with less ash had significantly fewer epiphytic colonies than did needles with large amounts of ash.

Cells of *Hormonema* are embedded in an extracellular gelatinous matrix. In the absence of ash, these epiphytic colonies were primarily restricted to the midrib depression on the adaxial surface of Douglas fir needles (Bernstein and Carroll 1977, Carroll 1979). Colonies presumably were not found on the lamina of needles, because this surface had fewer establishment sites and a less favorable microclimate (Bernstein and Carroll 1977). When ash was deposited on the needle surface, the microhabitat and climate were quickly altered. The ash could hold moisture for a longer period than the naked needle and contained many essential micronutrients (Fruchter et al. 1980) that were available to the fungi for growth. The small number of colonies in the midrib groove extended over the uncolonized needle surface and formed new colonies. These colonies in turn produced more new colonies. As a result, the fungal colonies became distributed over the entire adaxial surface of the ash-laden needle. Because ash adhered to the gelatinous slime of the colonies, more and more ash became stabilized on the surface of the needle. The growth of hyphae of other epiphytic fungi among the tephra further stabilized the ash layer on the needle.

Since the abaxial surface of Douglas fir needles had more establishment sites and a more moderate microclimate, more epiphytic fungal colonies were normally found on that surface than on the adaxial one (Bernstein and Carroll 1977, Carroll 1979). With ash-laden needles, the abaxial fungi existed not as individual colonies but as a continuous mat of hyphae over the stomatal surface.

Ash deposits were shown to have altered the habitats of other fungi (Cook et al. 1981). The moist ash layer on the soil presumably provided ideal conditions for the germination, growth, and fruiting of the phytopathogenic fungi *Sclerotinia*, *Phoma*, and *Cercosporella*.

The exploitation of the new ash habitat by epiphytic fungi may have had profound effects on conifer productivity. Phyllosphere fungi derive their carbon from leaf surfaces. In fact, certain surface yeasts have been shown to utilize the cuticular waxes of needles (McBride 1972). The persistent ash layer may have reduced light intensities and interferred with photosynthetic gas exchange. Reduced photosynthetic capabilities and the removal of carbon by the epiphytic fungi may have stressed and weakened the needles to such a degree that future production of wood, cones, and new needles will be affected.

Nonstructural carbohydrates of needles were significantly reduced 8 weeks after Douglas fir seedlings were artificially coated with volcanic ash. Presumably the ash on the surface of the needles affected photosynthesis by reducing light intensities or CO_2 concentration by altering stomatal function. Photosynthetic rates of ash-coated seedlings, however, were actually greater than those of the control plants, whereas dark respiration rates, compensation points, and light saturation levels were similar for both the ash-coated and control seedlings.

This phenomenon of enhanced photosynthetic rates in certain conifers has previously been reported by Bourdeau and Laverick (1958). When seedlings of white pine (*Pinus strobus*) and red pine (*Pinus resinosa*) were grown in full light, moderate shade, or deep shade and then experimentally subjected to light of varying intensities, those seedlings grown in moderate or deep shade achieved significantly higher rates of photosynthesis than those grown in full light. However, the growth of the seedlings from less than full light was significantly retarded. Apparently Douglas fir seedlings of this study responded to an artificial coat of ash in a similar manner as these shade-intolerant conifers did to experimental shading. In all likelihood, the coating of ash on the Douglas fir seedlings shaded the needles and reduced the light necessary for photosynthesis and carbohydrate accumulation. In nature, the surfaces of Douglas fir needles were coated with thick layers of ash and fungi that not only reduced light intensities but also acted as a barrier against the diffusion of CO_2 required for photosynthesis.

LITERATURE CITED

ATKINS, E. L., L. D. ANDERSON, and T. O. TUFT. 1954. Equipment and technique used in laboratory evaluation of pesticide dusts in toxicological studies with honeybees. *J. Econ. Entomol.* 47:965–969.

BERNSTEIN, M. E., and G. C. CARROLL. 1977. Microbial populations on Douglas fir needle surfaces. *Microbial Ecol.* 4:41–52.

BOURDEAU, P. F., and M. L. LAVERICK. 1958. Tolerance and photosynthetic adaptability to light intensity in white pine, red pine, hemlock and ailanthus seedlings. *Forest Sci.* 4:196–207.

CARROLL, G. C. 1979. Needle microepiphytes in a Douglas fir canopy: Biomass and distribution patterns. *Can. J. Botany* 57:1000–1007.

COOK, R. J., J. C. BARRON, R. I. PAPENDICK, and G. L. WILLIAMS III. 1981. Impact on agriculture of the Mount St. Helens eruptions. *Science* 221:16–22.

DANIELSEN, E. F. 1981. Trajectories of the Mount St. Helens eruption plume. *Science* 211:819–821.

daSILVEIRA, A. J., F. F. FEITOSA TELES, and J. W. STULL. 1978. A rapid technique for total nonstructural carbohydrate determination of plant tissue. *J. Agri. Food Chem.* 26:770–772.

del MORAL, R. 1981. Life returns to Mount St. Helens. *Nat. Hist.* 90:36–46.

FEDER, N., and T. P. O'BRIEN. 1968. Plant microtechnique: Some principles and new methods. *Am. J. Bot.* 55:123–142.

FINDLEY, R. 1981. Mount St. Helens: III. The day the sky fell. *Nat. Geographic* 159:50–65.

FISHER, D. B. 1968. Protein staining of ribboned epon sections for light microscopy. *Histochemie* 17:92–96.

FRUCHTER, J. S., D. E. ROBERTSON, J. C. EVANS, K. B. OLSEN, E. A. LEPEL, J. C. LAUL, K. H. ABEL, R. W. SANDERS, P. O. JACKSON, N. S. WOGMAN, R. W. PERKINS, H. H. VAN TUYL, R. H. BEAUCHAMP, J. W. SHADE, J. L. DANIEL, R. L. ERICKSON, G. A. SEHMEL, R. N. LEE, A. V. ROBINSON, O. R. MOSS, J. K. BRIANT, and W. C. CANNON. 1980. Mount St. Helens ash from the 18 May 1980 eruption: Chemical, physical, mineralogical, and biological properties. *Science* 209:1116−1125.

HAMMOND, P. E. 1980. Mount St. Helens: Portland State University professor describes Mount St. Helens volcanic activity. *Assoc. Eng. Geol. Newsletter* 23(3):12−24.

KÜCHLER, A. W. 1964. Potential natural vegetation of the conterminous United States. *Amer. Geograph. Soc. Spec. Pub. 36* 165 pp.

MCBRIDE, R. P. 1972. Larch leaf waxes utilized by *Sporobolomyces roseus in situ. Trans. Br. Mycol. Soc.* 58:329−331.

ROSENFELD, C. L. 1980. Observations on the Mount St. Helens eruption. *Am. Scient.* 68:494−509.

TABLE 12.1
DISTANCE FROM MOUNT ST. HELENS, ELEVATION,
AND VEGETATION TYPE OF SAMPLING SITES

Site	Distance from Mount St. Helens (km)	Elevation (m)	Vegetation type[a]
1	70	682.8	Fir–Hemlock
2	86	1268.0	Fir–Hemlock
3	102	737.6	Western Ponderosa
4	102	1036.3	Fir–Hemlock
5	113	591.3	Western Ponderosa
6	114	877.8	Western Ponderosa
7	119	609.6	Western Ponderosa
8	272	396.2	Sagebrush Steppe
9	305	554.7	Sagebrush Steppe
10	426	787.0	Douglas Fir
11	434	658.4	Douglas Fir
12	435	662.0	Douglas Fir
13	435	865.6	Western Ponderosa
14	454	853.4	Cedar–Hemlock–Pine
15	456	877.8	Cedar–Hemlock–Pine
16	456	894.6	Cedar–Hemlock–Pine
17	460	841.6	Cedar–Hemlock–Pine
18	460	359.7	Western Ponderosa
19	498	426.7	Western Ponderosa
20	523	1477.6	Western Spruce–Fir
21	580	853.4	Western Ponderosa
22	644	1073.8	Western Ponderosa

[a]After Küchler (1964).

TABLE 12.2
U.S. DEPARTMENT OF COMMERCE PRECIPITATION DATA FOR THE
PERIOD MAY 18 TO OCTOBER 10, 1980

Location	Total precipitation (cm)	Number of Days		
		>0.25 cm	>1.27 cm	>2.54 cm
Washington				
Packwood	17.30	19	4	0
Yakima	6.76	8	0	0
Ritzville	10.85	11	3	0
Spokane	11.20	10	2	0
Idaho				
Coeur d'Alene	20.93	19	7	0
Tensed	17.55	17	5	0
St. Maries	24.31	12	7	4
Elk River	28.37	29	5	0
Montana				
Haugan	31.14	27	9	2
Superior	23.75	27	4	2
Missoula	22.58	21	3	2

TABLE 12.3
THE AMOUNT OF NONSTRUCTURAL CARBOHYDRATES AS GLUCOSE
EQUIVALENTS IN ARTIFICIALLY ASH-COATED AND CONTROL SEEDLINGS
OF DOUGLAS FIR AFTER 3 AND 8 WEEKS OF TREATMENT

Treatment	Glucose/needle dry weight (mg/mg)
Three weeks	
Control needles	0.053 ± 0.009[a]
Ash-coated needles	0.043 ± 0.005
Eight weeks	
Control needles	0.039 ± 0.003
Ash-coated needles	0.026 ± 0.004

[a]95% confidence limit.

13

The Effect of Mount St. Helens Ash on the Development and Mortality of the Western Spruce Budworm and Other Insects*

Jerry J. Bromenshenk, R. C. Postle, G. M. Yamasaki, D. G. Fellin, and H. E. Reinhardt

ABSTRACT

Larvae of the western spruce budworm, *Choristoneura occidentalis* (Lepidoptera: Tortricidae), were fed volcanic ash from Mount St. Helens. Ingestion of ash significantly altered rates of development, survival, and the influence of internal parasites. Each effect differed depending on the method of administration of ash—through artificial diets containing ash or through ash-dusted foliage. The rate of development was significantly increased when larvae were fed ash mixed in artificial diets, and the size of pupae and adults, mainly males, was reduced. In addition, the ash appeared to have a prophylactic effect on a gut protozoan, *Nosema bombyx*. Budworm mortality, however, was not influenced by ash even when artificial diets contained 50% ash. Conversely, when stage $4-6$ larvae were fed ash-dusted foliage, mortality significantly increased. This response occurred for all treatments except those employing very light dustings of ash. Also, heavy coatings of ash on foliage reduced the number of larvae killed by parasitoids and occasionally produced dwarf adults. These findings are discussed with respect to a review of reported responses of other insects and with regard to implications for the design of toxicological bioassays.

INTRODUCTION

Insect mortality resulting from volcanic ash has been reported for eruptions of Katmai (Griggs 1919), Krakatau (Dammerman 1948), Paricutin (Segerstrom 1956), and Irazu (Wille and Fuentes 1975) and for

*This work was supported by USDA Cooperative Agreement INT-80-105-CA.

ashfalls in Java (Gennardus and Balvers 1983). Although large numbers of bees were killed by volcanic ash from Irazu volcano in Costa Rica, some pest species flourished, especially insects protected by waxy secretions. Elimination of natural control agents also seemed to be a cause of the increase of these species (Wille and Fuentes 1975). Similarly, Gennardus and Balvers (1984) reported an explosion of Lepidoptera, both in numbers of species and individuals, following prolonged volcanic activity and a long dry season in Java.

Several studies of the effects of ash from Mount St. Helens on insects were initiated following its explosive eruption on May 18, 1980 (Akre 1980). These investigations have documented a variety of responses by insects exposed to the ash. Because of their small size, their susceptibility to water loss resulting from abrasion and absorption, and the likelihood of being physically covered by ash, insects probably were impacted more than other forms of life by ash deposits from the eruption. Insect mortality has been attributed primarily to physical rather than chemical properties of the ash.

LABORATORY STUDIES

Laboratory and cage studies demonstrated mortality related to dessication in several species of bees (Johansen et al. 1981). Philogene (1972) reported that cockroaches (*Periplaneta americana*) died within 48 hr after walking on ash. House crickets (*Acheta domesticus*) placed on a 1-mm-deep layer of ash with ash-coated lettuce as a food source sustained 100% mortality within 40 hr. Juvenile insects were more acutely affected than adults. The cause of death was thought to be dessication due to cuticular abrasion combined with excessive release of saliva during prolonged grooming activities. Occlusion of the spiracular valves and accumulation of glutinous boli of ash and food in the gut were contributory factors (Edwards and Schwartz 1981).

Physiological experiments with representatives of Diptera, Hymenoptera, Orthoptera, Lepidoptera, and Coleoptera suggested that predaceous and parasitic insects (especially Hymenoptera) were particularly susceptible to harm from ash. Ash on insect bodies affected the ability of insects such as honey bees (*Apis mellifera*) to fly. Adult house flies (*Musca domestica*) and brown-banded cockroaches (*Supella longipalpa*) exposed to a 1-cm-deep layer of ash lost 30–40% of their live body weight after 8 hr, whereas the wax moth (*Galleria mellonella*) and the yellow mealworm (*Tenebrio molitor*) sustained a <5% loss. Mealworms appear to have evolved an integument that offers protection in dusty environments, whereas immature wax moths avoided contact with ash by spinning cocoons. Entrapment, abrasion of epicuticular waxes and, to a lesser extent, absorption of water by ash were responsible for observed mortalities. Ingestion and inhalation of ash particles were not thought to be major mortality factors (Brown and bin Hussain 1981).

Ash on foliage produced limited mortality (14–18%) in adult codling moths (*Cydia pmomonella*) and no ovicidal effect but a high mortality (78%) in neonate larvae (Howell 1981). Ash-coated leaves inhibited feeding and caused adult mortality of *Otiorhynchus* root weevils. Constant body contact with ash was fatal, partly because of wounding of the membranous parts of the body by fine, sharp particles in the ash (Shanks and Chase 1981). Ash applied to leafy pear shoots reduced oviposition, induced mortality of eggs and nymphs, and reduced longevity and oviposition rates of pear psylla (*Psylla pyricola*). These tests indicated that in one generation a 70–80% reduction of pear psylla could occur (Fye 1983).

FIELD STUDIES

Field surveys support these laboratory observations. Foraging ants, bees, and yellow jackets sustained severe initial losses in areas of heavy ash fallout. Ants of the genus *Formica* rapidly recovered, especially after rain had dampened the ash so that the ants could walk on the surface and remain relatively dust free. Yellow Jacket populations were devastated, probably because of a combination of the initial ashfall, wind-blown ash, and cold, wet weather on aerial nests (Akre et al. 1981).

Field forces of honey bees were annihilated in the heavy ashfall region of eastern Washington. Breaks in brood rearing (cessation of egg laying) of 5–10 days were reported by several beekeepers. Laboratory tests indicated that exposure to ash could induce mortality in uncapped larvae (Johansen et al. 1981). Johansen (1980) estimated 12,000 colonies were destroyed or severely damaged in the Columbia River basin. Later observations confirmed a total loss of 1500 colonies by the autumn of 1980. The remaining colonies sustained an average loss of 50,000 workers per colony during the 1980 season. Honey yields of colonies remaining in the heavy ashfall region were reduced by 13.6–22.7 kg per colony. Colonies moved to cleaner areas lost about 4.5 kg, whereas ~300,000 colonies in moderate ash areas did not display measurable reductions in honey, in spite of a loss of about 20,000 workers during the initial dustfall (Johansen et al. 1981).

Losses of honey, bee pollinators, and related sources of income were estimated at $1–2 million (Johansen 1980, Cook et al. 1981). The impact on bee-pollinated crops was partly mitigated because pollination of fruit trees in the main fallout area had been completed before the eruption. An exceptionally wet spring with fewer than normal hot days in July and August provided better than average nectar flows (Johansen et al. 1981).

Other pollinators were also affected. The ash affected foraging distance and shortened nesting activity of female leafcutting bees (*Megachile rotundata*). Growers reduced the impact by delaying emergence of bees and by ash clean-up activities near bee shelters. Alkali bees (*Nomia melanderi*) seemed relatively unaffected, probably because of adaptations to soil nesting. Also, adequate humidity facilitated removal of ash from body surfaces. Early emerging bumblebee queens were unsuccessful in starting colonies in areas subjected to large amounts of ash, but late emerging individuals and species survived with little apparent effect (Johansen et al. 1981). Bumblebees seemed able to effectively clean ash from their bodies and retained the ability to fly (unpublished work of Brown 1981, cited in Johansen et al. 1981). Klostermeyer et al. (1981) observed a populaton of orchard mason bees (*Osmia lignaria*) leave their nests the day after the ashfall, but none returned from foraging. Johansen et al. (1981), however, reported that ~25% of the mason bees had already produced progeny at the time of the eruption. Thus *Osmia* populations were probably reduced but not annihilated.

Suppression of orchard populations of pear psyllids through several instars following exposure was appreciable. In addition, populations of winged predators were decimated, although populations of *Deraeocoris brevis* recovered because eggs were laid in leaf tissue before the eruption (Fye 1983). Arthropod populations in wheat fields apparently were not severely affected. The ash impact was described as being similar to a nonpersistent, broad spectrum insecticide, causing a temporary reduction in numbers of some species. Rain 4 days after the heaviest ashfall settled the dust and was thought to have reduced adverse effects on insects (Klostermeyer et al. 1981). Scheduled spraying of 120,000 acres of rangeland for grasshopper control was abandoned because of eradication of grasshoppers by ash and subsequent cool, rainy conditions (Kiem 1981, personal communication to Cook, cited in Cook et al. 1981). In potato fields, adult Colorado potato beetles (*Leptinotarsa decemlineata*) died after 24 hr of exposure to ash. First, second, and third instar beetles were more sensitive than adults, while fourth instar were more resistant. Most of the beetle populations survived in plots that were "washed" by overhead irrigation (unpublished work by Powell 1981, cited by Cook et al. 1981).

EFFECTS ON WESTERN SPRUCE BUDWORM

As much as 76 mm of ash fell on Moses Lake and Ritzville, Washington, whereas in the forested area of western Montana, ash accumulated to depths of 1.6–25.4 mm (Moen and McLucas 1981). The 4-day period from May 18–21, when ash fallout was the greatest, coincided with the peak period of larval emergence and dispersal of western spruce budworm from their overwintering sites. Shortly thereafter, western spruce budworm larvae began their annual period of feeding and growth. The effect the ash deposited on coniferous foliage had on feeding spruce budworm and subsequent impacts on budworm populations in these forests were of interest to resource managers and others administering forests in the Pacific Northwest.

METHODS AND PROCEDURES

Our objectives were to determine the toxicity of volcanic ash to western spruce budworm larvae and to examine responses resulting from ingestion of ash contained in an artificial diet compared with ash on foliage.

OBTAINING LARVAE AND ASH

In April 1981, 100 bole sections were cut from seven Douglas fir trees growing in a spruce budworm-infested area of the Deerlodge National Forest near Pipestone, Montana. The tree sections were transported to Missoula, Montana, and temporarily stored in a walk-in cooler at a temperature of 5°C.

Beginning in May, boles were brought into the laboratory and placed into 5-gal cardboard ice cream cartons. Two glass vials affixed to the lid of each container served as emergence traps. With the laboratory kept at room temperature and continuously lighted, larvae overwintering under bark flakes on the bole sections emerged, moving toward the light and into the vials, where they were collected. More than 15,000 larvae were obtained in this manner, the emergence per tree ranging from 238 to 1268 larvae per square meter of bark surface (L. J. Theroux, personal communication 1981).

Ash for these tests was acquired from a landfill near St. Maries, Idaho, where it had been piled after removal from city streets. Ash was taken from the center of the piles since the outer several centimeters had visibly weathered. The ash that fell near St. Maries and in other areas of Idaho, eastern Washington, and western Montana was quite abrasive. Particle sizes ranged from <2 μm (2−6 wt. %) to 50 μm (22−88 wt. %). The ash contained small amounts of such toxic elements as arsenic, mercury, lead, cadmium, and fluoride (Fruchter et al. 1980, Moen and McLucas 1981).

APPLICATIONS OF THE TREATMENTS

Larvae Fed Foliage Dusted with Ash

A vacuum chamber (Farrar et al. 1948, Atkins et al. 1954, Atkins et al. 1973) was used to apply volcanic ash in controlled amounts to foliage. Foliage to be dusted (terminal branchlets, 15−25 cm in length and freshly cut from mature Douglas fir trees) was placed vertically in a bell jar−shaped chamber. A weighed amount of ash was placed on a watch glass suspended in the upper part of the chamber. In use, the chamber was evacuated to 63.5 mm Hg; then an electrically controlled solenoid valve opened to admit entry of air from outside of the chamber. The inrushing air imploded and dispersed the ash.

This process produced a uniform coating of dust to all surfaces of needles and stems, with just slightly more dust on the adaxial surfaces. This is similar to the way ash was deposited on coniferous foliage by Mount St. Helens.

The amount of ash delivered was determined using filter paper targets. On the basis of this information (Fig. 13.1), we chose eight independent dusting levels of 0, 2, 4, 8, 12, 16, 24, and 32 g. The upper limit of 32 g reflects the holding capacity of foliage for ash and approximates the maximum amount found adhering to field-collected Douglas fir foliage 5 months after the May, 1980, eruption (Bilderback and Slone, Chapter 12, this volume).

Using this apparatus, we were able to simultaneously dust 2 sets of 12 branchlets at the same or different dosages. Once dusted, the foliage was placed in Randall cages on racks with the stems immersed in water.

Fourth stage larvae were selected at random and five were placed on the foliage in each cage. These larvae had been reared on an artificial diet (Lyan et al. 1972) from stage 2 larvae that had emerged from the bole sections. Once each week, the foliage in the cage was examined and the number of larvae alive and feeding were counted, along with the number dead and the number infected by parasites.

This experiment was designed as a randomized block with 10 cages for each of the 8 dosages. Blocks of 80 cages were replicated 8 times. Cages in each block were randomly distributed on racks in a

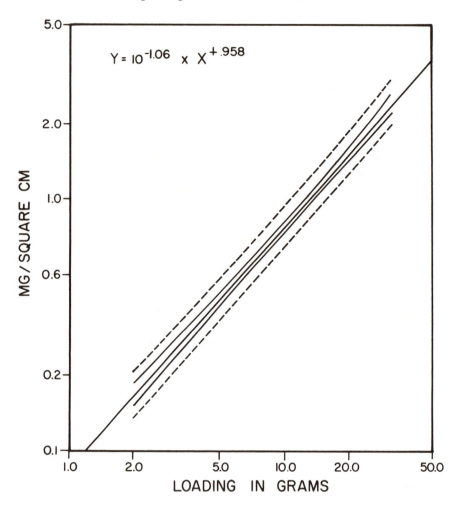

FIG. 13.1   Calibration curve for dusting chamber; surface deposition of ash results from different application rates. Ash application in mg/cm².

greenhouse, where the temperature, light, and humidity were regulated to approximate natural diurnal fluctuations. Humidity ranged from 16 to 96% and temperature from 11.1 to 34.4°C.

Larvae Fed Artificial Diet Containing Ash
Volcanic ash was mixed with water to make a slurry before being mixed with other ingredients of an artificial diet regularly used by many budworm researchers (Lyan et al. 1972). We tested seven different

preparations containing 0, 8, 17.5, 25, 30, 40, and 50%, respectively, of ash on a dry weight basis. The higher dosages exceeded the highest concentration of ash per gram of foliage observed on conifers in forests where ash fell following the eruption (Bilderback and Slone, Chapter 12, this volume).

Larvae were individually reared on the ash-containing diet in disposable 5-cm tissue culture dishes; the diet was replaced as needed. Three trials were conducted, each consisting of four treatment levels of ash with 50 larvae per treatment. In the first trial, 0, 8, 17.5, and 25% portions of the diet media were replaced with ash; in the second and third trials, 0, 30, 40, and 50% portions of the media were replaced with ash.

All larval rearing was done in a windowless laboratory room. Although temperature and humidity were not strictly controlled, forced air cooling was used to keep maximum temperatures below 25°C. Each budworm was examined daily to determine the stage of development, as well as the time of larval pupation and emergence of adults. The number of live, dead, and parasitized larvae were also recorded while pupae were sexed and weighed.

STATISTICS

The hypothesis being tested for dusted foliage was that different levels of ash had no effect on mortality of western spruce budworm. The hypothesis being tested for ash in the artificial diet was essentially the same as for dusted foliage. In addition, we tested whether ash had any effect on development rate; this was indicated by the time from instar to instar for different levels of ash, as well as by differences between mean number of days needed for transition to each instar and measurements of these differences using several linear regression models and analysis of variance techniques (Chatterjee and Prie 1977, Ostile 1971, Searle 1971, Snedecor 1971).

Before analyses were run, methods and data were checked to make certain they met assumptions of linearity, additivity, independence, and homogeneity of variance (Bartlett's test). The results of analysis of variance were evaluated using Tukey's multiple range test, 95% confidence intervals, or both.

RESULTS

Prior to conducting the feeding tests, observations were made of the behavior of larvae exposed to ash-covered surfaces. Using the vacuum chamber, thin coatings of ash were applied to the surfaces of petri dishes; then half of each dish was wiped free of ash.

When placed in the clean portion of the dish, third-stage larvae avoided contact with the ash, usually turning away whenever they encountered the material. Larvae placed in the ash-covered half of the dish usually began to crawl about immediately. Those that reached the ash-free portion of the dish seldom crawled back into the ash-covered section. Larvae that did not crawl away from the ash-covered surface generally ceased moving within a matter of a few minutes and died.

EXPERIMENTS WITH ASH-DUSTED FOLIAGE

Larval survival diminished significantly with time and with increased concentrations of ash (Fig. 13.2; Table 13.1). Application rates of 8 g or more of ash to foliage produced significant reductions in survival. Only a slight interaction was detected between the periods of observation and the concentration of ash. Ash application rates in excess of 16 g significantly lowered the percentage of larvae killed by parasitoids (Table 13.2). Mean percentage of mortality attributed to ash ranged from 6 to 18% for the various dose levels, whereas mortality caused by parasitoids ranged from 16 to 30%. Ash dosages of 12 g or more significantly suppressed parasitoid populations. Occasionally, adults of budworm reared on ash-laden foliage were dwarfed, some being <50% normal size.

EXPERIMENTS WITH ASH-CONTAINING DIET

We were unable to detect any differences in survival of larvae related to the concentrations of volcanic ash mixed into the artificial diet. In feeding trials in which 8, 17.5, and 25% portions of the diet were composed of ash, larval survival tended to be higher than for the control diet containing no ash (Fig. 13.3). Examination of the data revealed that a considerable proportion of the control larvae were killed by the protozoan *Nosema bombycis*.

In later feeding trials in which 30, 40, and 50% portions of the diet were composed of ash, the diet was treated with Benomyl® to kill the protozoa (J. L. Robertson, personal communication 1981). In these trials, where the diet was treated prophylactically to suppress the protozoan, there was no treatment effect.

Volcanic ash in the diet significantly affected development rate by altering the lengths of the larval stages, particularly the sixth stage. For all feeding tests in which diets contained ash, the length of time larvae were in stage 6 was shorter than in trials where no ash had been added to the diet (Fig. 13.4; Table 13.3).

Ingestion of ash by larvae also had effects on the size of adults that developed from surviving larvae. The mean weight of males tended to decline in treatments where larvae were fed an artificial diet with higher percentages of ash. The weight of males ranged from 0.1 g in the controls to 0.067 and 0.068 g in treatments with 30 and 40% ash, respectively. With one exception, the weight of females was not influenced by the amount of ash ingested by larvae; the mean weights of females ranged from 0.103 to 0.107 g regardless of treatment. In one replication, however, weight of females in the control group averaged 0.114 g while those developing from larvae on diets with 50% ash weighed 0.074 g.

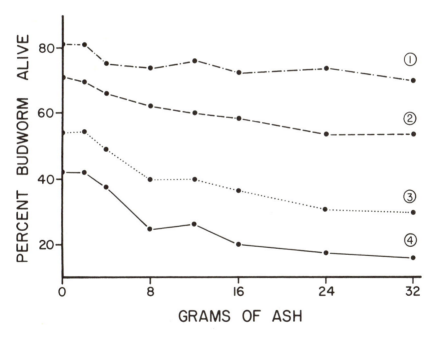

FIG. 13.2   Effects of ingesting ash-dusted foliage on survival of western spruce budworm. Observations were made on a weekly basis. Observation 1 represents the initial (first week) count, and observation 4 the final count when the cages were disassembled and carefully searched for living larvae.

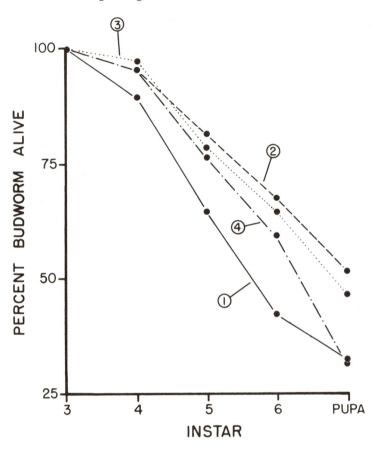

FIG. 13.3 Survival of western spruce budworm larvae fed varying concentrations of volcanic ash in an artificial diet. All populations were infected with the protozoan *Nosema bombycis*. Ash concentrations were 0, 8, 17.5, and 25% as indicated by 1–4, respectively.

DISCUSSION

On the basis of the combined results of our tests, it appears that the ashfall from the May 18, 1980, eruption of Mount St. Helens most likely did have a measurable impact on western spruce budworm populations. A comparison of the effect of ash on mortality of larvae in the laboratory with the concentrations of ash actually deposited on trees in forests suggests that feeding on coated foliage could have

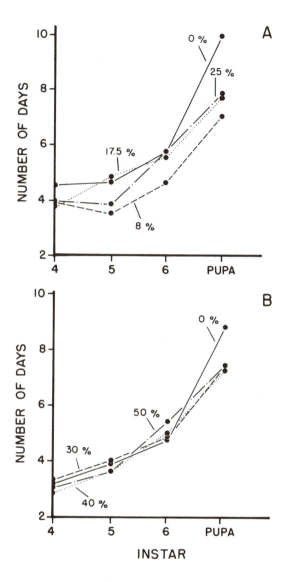

FIG. 13.4   Length of developmental periods for
spruce budworm fed varying concentrations of ash in
an artificial diet. (A) Ash concentrations of 0–25%;
all populations were infected by *Nosema bombycis*. (B)
Ash concentrations of 0–50%; the diet was treated
prophylactically to suppress *Nosema bombycis*. Ash levels
1–4 same as Fig. 13.3.

reduced populations in the forest by 15–20% (corrected for reductions in parasitoid populations).

The initial effect of the eruption was apparently on stage 2 larvae, which were at the peak of their emergence from overwintering sites and preparing to feed. They were most likely to be subjected to direct contact with the ash and were particularly vulnerable to entrapment, abrasion, and dessication. Many larvae surviving the initial ash fallout no doubt died later after feeding on ash-covered foliage. As a result of larvae being killed by the ash, the incidence of defoliation was probably reduced. Furthermore, behavioral observations recorded that larvae consumed less foliage with increased concentrations of ash.

Our findings of ash-induced mortality and suppression of natural control agents are supported by other research, reviewed in the introduction to this chapter. Our results are not totally corroborated by other investigators. For example, *Manduca sexta* were force-fed artificial diets containing 5% water substituted by a variety of inert materials including volcanic ash, but they did not demonstrate inhibition of larval development with respect to larval weight, duration of penultimate instar, or days required to pupate (Brown and bin Hussain 1981). Our findings of shortened development periods and reduced male pupal weight for budworm that were fed ash substitutes for part of their diet differ from those for *Manduca*. The discrepancies in the two studies may simply be the result of differences in species response or because we used a broader range of ash concentrations, larger sample sizes, more experimental replications, longer periods of exposure of larvae, and greenhouse facilities for rearing rather than controlled environmental chambers.

We believe that the disparate results obtained by feeding larvae on ash-covered foliage compared to larvae fed ash incorporated into an artificial diet do not result from the range of concentrations of ash used. At the maximum dosages, dust was applied to the foliage in concentrations equivalent to ∼40% ash to dry weight of needles, whereas ash composed 50% by weight of the artificial diet.

It is likely that other factors could explain these differences:

1. Abrasion of the larval cuticle, and subsequent water loss, may have been a major mortality factor of larvae feeding on the ash-dusted foliage. The feeding trials using ash in an artificial diet indicate that ash is relatively nontoxic when ingested by feeding larvae, although physical or chemical properties of the volcanic ash may have been altered in the process of incorporating the ash with the other ingredients in the artificial diet.

2. Conversely, physiological or biochemical processes in foliage may have been affected by the ash, a circumstance that would not be reproducible by adding ash to an artificial diet.

3. The increased developmental rate observed in larvae reared on the artificial diet suggests that the ash, rather than having a toxic effect, may have elicited a nutritive or feeding stimulus response in the larvae. Inhibition of feeding by larvae on ash-dusted foliage, however, does not support the hypothesis of a feeding stimulant.

4. Starvation, or some other form of stress response, is suggested by the development of occasional dwarf adults and reduced pupal weight combined with the increased developmental rate of larvae feeding on the artificial diet. Decreased larval survival, reduced egg production, and lighter pupal weights were clearly demonstrated for the large aspen tortrix (*Choristoneura conflictana*) when larvae were subjected to starvation regimes and different host plants. Larvae fed on alder took more than twice as long to develop as those fed on aspen (Beckwith 1970).

Volcanic ash, when ingested by larvae feeding either on ash-covered foliage or on an artificial diet containing ash, reduces or inhibits the action of internal larval parasitoids. The effect of ash on these internal parasites may be more significant than the direct effect of ash on the larvae that harbor them. Our larval mortality data indicate that ingestion of volcanic ash by larvae could reduce the effectiveness of internal parasites as a control agent of western spruce budworm larvae by >50%. Thus much of the natural insecticidal potential of ash for budworms may be offset by suppression of natural control agents.

Whatever the mechanisms involved in the response of western spruce budworm larvae or its natural enemies to the introduction of volcanic ash into their diets, our investigations demonstrate the potential for misinterpretation in assessing the effects of any factor on insect population dynamics on the basis of the results of a single type of treatment or limited sample sizes. Investigators in many laboratories routinely conduct toxicological bioassays of chemical insecticidal formulations using laboratory populations reared on artificial diets in controlled environments. Researchers have been cautioned about extrapolation to field conditions based solely on a single type of laboratory bioassay (Robertson and Haverty 1981). Our research demonstrates the potential for error when results are based on small sample sizes, short observation periods, or the use of artificial diets versus more "natural" foliage-type diets. The two techniques produced dramatically different results.

LITERATURE CITED

AKRE, R. D. 1980. Effect of volcanic ash on insects. In *Aftermath of Mount St. Helens*, Washington State Univ. Conference, July 8–9, pp. 14–15.

AKRE, R. D., L. D. HANSEN, H. C. REED, and L. D. CORPUS. 1981. Effects of ash from Mt. St. Helens on ants and yellowjackets. *Melandria* 37:1–19.

ATKINS, E. L., L. D. ANDERSON, and J. TUFT. 1954. Equipment and technique used in laboratory evaluation of pesticide dusts in toxicological studies with honey bees. *Econ. Entomol.* 47(6):965–969.

ATKINS, E. L., E. A. Greenwood, and R. L. MacDonald. 1973. *Toxicity of Pesticides and Other Agricultural Chemicals to Honey Bees.* Laboratory Study, Univ. California Agricultural Extension M-16, (Revised 9/73), 38 pp.

BECKWITH, R. C. 1970. Influence of host on larval survival and adult fecundity of *Choristoneura conflictana* (Lepidoptera: Tortricidae). *Can. Entomol.* 102:1474–1480.

BROWN, J. J., and Y. bin HUSSAIN. 1981. Physiological effects of volcanic ash upon selected insects in the laboratory. *Melandria* 37:30–38.

CHATTERJEE, S., and B. PRIE. 1977. *Regression Analysis by Example.* John Wiley, New York, 228 pp.

COOK, R. J., J. C. BARRON, R. I. PAPPENDICK, and G. J. WILLIAMS III. 1981. Impact on agriculture of the Mount St. Helens eruptions. *Science* 211:16–22.

DAMMERMAN, K. W. 1948. The Fauna of Krakatau 1883–1933. Verhandelingen Der Koninklijke Nederlandsche Akademie Van Wetenschappen, AFD. Natuurkunde Tweede Sectie. *K. Ned. Akad. Wet. Versl. Gewone Vergad. Afd. Natuurkd.* 44:1–594.

EDWARDS, J. S., and L. M. SCHWARTZ. 1981. Mount St. Helens ash: A natural insecticide. *Can. J. Entomol.* 59(4):714–715.

FARRAR, M. D., W. C. KANE, and H. W. SMITH. 1948. Vacuum dusting of insects and plants. *J. Econ. Entomol.* 41(4):647–648.

FYE, R. E. 1983. Impact of volcanic ash on pear psylla (Homoptera: Psyllidae) and associated predators. *Environ. Entomol.* 12(1):222–226.

FRUCHTER, J. S., D. E. ROBERTSON, J. C. EVANS, K. B. OLSEN, E. A. LEPEL, J. C. LAUL, K. H. ABEL, R. W. SANDERS, P. O. JACKSON, N. S. WOGMAN, R. W. PERKINS, H. H. VAN TUYL, R. H. BEAUCHAMP, J. W. SHADE, J. L. DANIEL, R. L. ERICKSON, G. A. SEHMEL, R. N. LEE, A. V. ROBINSON, O. R. MOSS, J. K. BRIANT, and W. C. CANNON. 1980. Mount St. Helens Ash from the 18 May 1980 eruption: Chemical, physical, mineralogical, and biological properties. *Science* 209:1116–1125.

GENNARDUS, B., and L. F. BALVERS. 1983. Observations of various cultures of caterpillars and butterflies following a rain of volcanic ash in Java, Indonesia. *Entomol. Ber.* 43(5):69–71.

———. 1984. Further observations on butterflies after a volcanic ash rain. *Entomol. Ber.* 44(1):2–4.

GRIGGS, R. F. 1919. Scientific results of the Katmai expedition of the National Geographic Society. IV. The character of the eruption as indicated by its effects on nearby vegetation. *Ohio J. Sci.* 19:173–209.

HOWELL, J. F. 1981. Codling moth: The effects of volcanic ash from the eruption of Mount St. Helens on egg, larval, and adult survival. *Melandria* 37:50–55.

JOHANSEN, C. A. 1980. Mount Saint Helens blows the season for many Washington beekeepers. *Am. Bee J.* 120:500–502.

JOHANSEN, C. A., J. D. EVES, D. F. MAYER, J. C. BACH, M. E. NEDROW, and C. W. KIOUS. 1981. Effects of ash from Mt. St. Helens on bees. *Melandria* 37:20–29.

KLOSTERMEYER, E. C., L. D. CORPUS, and C. L. CAMPBELL. 1981. Population changes in arthropods in wheat following volcanic ash fall-out. *Melandria* 37:45–49.

LYAN, R. L., S. J. BROWN, and J. L. ROBERTSON. 1972. Contact toxicity of sixteen insecticides applied to forest tent caterpillars reared on artificial diet. *J. Econ. Entomol.* 65:928–930.

MOEN, W. S., and G. B. McLUCAS. 1981. Mount St. Helens ash properties and possible uses. *Washington Dept. Nat. Res. Rep. Invest.* 24. 60 pp.

OSTILE, B. 1971. *Statistics in Research* (2d Ed.). Iowa State Univ. Press, Ames. 370 pp.

PHILOGENE, B. J. R. 1972. Volcanic ash for insect control. *Can. Entomol.* 104(9):1487.

ROBERTSON, J. L., and M. H. HAVERTY. 1981. Multiphase laboratory bioassays to select chemicals for field testing on the western spruce budworm. *J. Econ. Entomol.* 74: 148–153.

SEARLE, S. R. 1971. *Linear Models.* John Wiley, New York, 532 pp.

SEGERSTROM, K. 1956. Erosion Studies at Paricutin, State of Michoacan, Mexico. *U.S. Geol. Surv. Bull. 965A,* 164 pp.

SHANKS, C. H., JR., and D. L. CHASE. 1981. Effect of volcanic ash on adult *Otiorhynchus.* (Coleoptera: Curculionidae). *Melandria* 37:63–66.

SNEDECOR, G. W. 1971. *Statistical Methods* (6th Ed.). Iowa State Univ. Press, Ames, 367 pp.

WILLE, A., and G. FUENTES. 1975. Effect of the ash from the volcano Irazu (Costa Rica) on some insects. *Rev. Biol. Trop.* 23(2):165–175.

TABLE 13.1
Analysis of Variance and Confidence Intervals for Percentage of Survival of Spruce Budworm Larvae Fed Ash-Dusted Foliage[a]

Source of variation	d.f.	SS	MS	F ratio	F probability
Treatment	31	99.07	3.20	65.33	0.000
Ash	7	12.77	1.82	37.29	0.000
Time	3	84.44	28.15	575.43	0.000
Ash × time	21	1.87	0.09	1.82	0.004
Error	2528	123.66	0.05	—	—
Total	2559	222.73			

Amount of ash (g)	N	Survival (%)	Standard deviation	95% confidence limits
0	320	0.6181	0.2608	0.5894, 0.6468
2	320	0.6138	0.2646	0.5847, 0.6429
4	320	0.5563	0.2726	0.5263, 0.5863
8	324	0.4975	0.2912	0.4656, 0.5294
12	320	0.5038	0.2945	0.4714, 0.5362
16	320	0.4663	0.3007	0.4332, 0.4994
24	320	0.4375	0.3109	0.4033, 0.4717
32	316	0.4196	0.2956	0.3868, 0.4524
Total	2560	0.5142	0.2950	0.5028, 0.5256

[a]d.f. = degrees of freedom; SS = sum of squares; MS = mean square.

TABLE 13.2

ANALYSIS OF VARIANCE AND CONFIDENCE INTERVALS FOR PERCENTAGE
OF PARASITISM OF SPRUCE BUDWORM LARVAE
FED ASH-DUSTED FOLIAGE[a]

Source of variation	d.f.	SS	MS	F ratio	F probability
Treatment	7	1.80	0.26	5.22	0.000
Error	632	31.13	0.04		
Total	639	32.93			

Amount of ash (g)	N	Parasitism (%)	Standard deviation	95% confidence limits
0	80	0.2975	0.0466	0.2871, 0.3079
2	80	0.2750	0.0500	0.2639, 0.2861
4	80	0.2950	0.0516	0.2835, 0.3065
8	81	0.2741	0.0564	0.2616, 0.2866
12	80	0.2525	0.0483	0.2418, 0.2632
16	80	0.1885	0.0446	0.1676, 0.1874
24	80	0.1825	0.0539	0.1705, 0.1945
32	79	0.1595	0.0424	0.1500, 0.1690
Total	640	0.2394	0.0515	0.2354, 0.2434

[a]d.f. = degrees of freedom; SS = sum of squares; MS = mean square.

TABLE 13.3

DAYS REQUIRED TO COMPLETE SIXTH INSTAR BY SPRUCE BUDWORM
LARVAE FED ASH IN AN ARTIFICIAL DIET[a]

Amount of ash (%)	N	Mean	Standard deviation	95% confidence limits
0	94	8.8830	3.8181	8.1009, 10.6651
0	30	9.9333	3.9994	8.4934, 11.3732
8	48	7.0417	1.9347	6.4921, 7.5913
17.5	46	7.7174	2.6303	6.9564, 8.4836
25	29	7.8276	2.8917	6.7716, 8.8884
30	82	7.5488	2.5587	6.9868, 8.1108
40	80	7.3250	2.7640	6.7109, 7.9390
50	100	7.4600	2.9074	6.8832, 8.0368

14

The Role of Vegetation in the Recycling and Redistribution of Volcanic Mercury

B. Z. Siegel and S. M. Siegel

ABSTRACT

Natural, nonmineralized sources consisting mainly of volcanic vents, fissures, fumeroles, and hot springs account for ~70% of the mercury released annually into the atmosphere. The 1977 Kilauea flank eruption ("Kalalua") is estimated to have released ~2.5 metric tons of mercury, and fallout at the Sulfur Bank fumeroles reaches 300 kg/km² annually. Weathering of basalt and lava also is a slow but continuous source. That plants play a significant role in the natural mercury cycle has only begun to be appreciated since the 1970s. Although soil does not accumulate mercury, most plants do; however, a significant number have a lower tissue mercury content than does the soil. The accumulative ability of plants will vary with species, immediate environment, and proximity to volcanic and other local or regional geological features. Included in this study are species of lichens, basidiomycetes, algae, bryophytes, and a wide variety of gymnosperms and angiosperms. Within this group, ~70% of those examined thus far also release mercury as a vapor. Mercury accumulation ratios (tissue:soil) may be as large as 90:1 or more, but normally fall in the two- to ten-fold range. Release of the element as vapor commonly occurs at rates of ~1.0 µg/kg/hr at 25°C, but in this study was observed to reach 39.5 µg/kg/hr in the garlic vine *Pseudocalymma*. By comparison, most soils tested at 25°C in the absence of plants released mercury at only 20−30% of the typical plant rate. The ability to accumulate mercury from the substratum and to release it as vapor into the atmosphere provides plants with a novel but potentially significant means for detoxification of mercuriferous soils. Plants not only may be indicators of local degassing associated with volcanism (Mount St. Helens), but also may be a major factor influencing, if not regulating, land and land−air mercury distribution patterns on a global scale.

INTRODUCTION

The significance of volcanic and fumerolic emissions as primary sources of environmental mercury has been well documented at Kilauea and Mauna Loa in Hawaii (Eshleman et al. 1971; Siegel et al. 1973, 1975; Siegel and Siegel 1975, 1978*a*, *b*, *c*; Connor 1979; Siegel and Siegel 1979). Comparative studies have also shown that the ambient mercury levels at Krysuvik, Myvatn, Geysir, and Mount Hekla in Iceland and Mount Erebus in Antarctica are of the same order of magnitude as those found in Hawaii. Commonly this concentration of mercury lies within the range of $10-25$ $\mu g/m^3$, which is ~30 times higher than found at nonthermal sites within volcanic regions and perhaps 2500 times higher than what can be considered as normal baseline values for eastern Pacific air (McCarthy et al. 1973). Continuous emissions from Kilauea and Mauna Loa on the island of Hawaii, which significantly increase during periods of eruption, deliver mercury to volcanically inactive sites including Oahu (Honolulu), ~400 km to the northwest (Siegel and Siegel 1978*a*). This atmospheric mercury adds substantially to the amount released locally by the weathering of igneous rocks.

Once it enters the soil and groundwater, mercury is first accumulated by plants and in many cases released again as mercury vapor into the atmosphere. This combined accumulative and revolatilization capability of plants constitutes a major link in the recycling of volcanic mercury into the atmosphere and, by the localized removal of soil mercury, an important factor in postvolcanic colonization.

MATERIALS AND METHODS

All mercury analyses of suitably collected, processed, and replicated samples were done by flameless atomic absorption (Siegel et al. 1975). Replicate soil or plant tissues were digested in nitric acid; air samples of at least 0.25 m^3 were collected in nitric acid—saturated fiberglass for total mercury or on gold foil for elemental mercury at pump flow rates averaging $2-3$ liters/m.

Volatile mercury released from leaves, other plant tissues, and soils were measured by suspending $4-5$ cm^2 of acid-cleaned gold foil over the samples in closed humidified containers at 25°C (Siegel et al. 1974, Kama and Siegel 1980, Siegel et al. 1980*a*).

RESULTS AND DISCUSSION

PRIMARY SOURCES OF MERCURY

Andren and Nriagu (1979) modified the mercury cycle proposed by Wollast et al. (1975) using 1973 data, and concluded that for the continental United States 32% of the atmospheric burden was of anthropogenic origin. They also adopted an estimate of 10^9 g/yr as the natural degassing rate for the same land surface. Without questioning the details of such models, clearly most of the atmospheric mercury burden originates in nature. In the United States, 471 metric tons of mercury were lost directly to the atmosphere from commercial industrial uses during 1973 (U.S. Environmental Protection Agency 1975; see also Joensuu 1971). Natural release by degassing, however, contributed ~1000 metric tons during the same year. One may assume that industrial Europe and Asia release similar amounts of mercury, and natural (nonanthropogenic) sources account for >70% of the atmospheric burden in much of Africa, agricultural east-central Europe, the steppes and highlands of central Asia, and the islands of the Pacific Ocean. Natural mercury arises from either the weathering of igneous rocks or direct degassing from volcanic vents, fumeroles, fissures, and other rift zone features (Table 14.1) and subsequent interactions with the soil surface (Table 14.2). Although freshly extruded lava from Kilauea may lose 90% of its 700 μg/kg of mercury in only 20 yr (Siegel and Siegel 1973), it is more probable that losses of such magnitude from igneous rocks require decades.

An idea of the amount of mercury potentially available from weathering of lavas in Hawaii can be gotten by using the example of the 1868 Kilauea East Rift eruption which totaled ~10^{12} kg of solids (Mac Donald and Abbott 1970). With the assumption of an average initial mercury content of 200 μg/kg (Table 14.1), lavas would introduce 200 metric tons of mercury into the environment, if weathered all at once. Even if averaged over a century, a single massive eruption could make a significant contribution. Over the past 200 yr, Mauna Loa and Kilauea have extruded ~20 times more magma than the 1868 eruption and should therefore contain a potential 4000 metric tons of mercury. Even if we reduce this estimate by an order of magnitude (to be conservative), the amount of 400 metric tons remains impressive. It cannot, however, be assumed that all of the mercury released by weathering of igneous rocks will enter the atmosphere. A portion of it will most likely remain in the silicate matrix, and a larger portion will be subjected to dissolution by volcanic acid rain (Kratky et al. 1974) and will enter the groundwater system in solution.

Aerometric studies for locations in Iceland, Hawaii, and Antarctica show that these separate and diverse areas produce similar amounts of mercury. Although concentrations of 35 elements have been determined in the Mount St. Helens plume of May 18, 1980 (Vossler et al. 1981), mercury was not included; hence we can only speculate that the volcanoes of the Cascades are also rich sources of mercury. There is ample evidence for mercuriferous thermal areas along western North America from California to Alaska, including Alberta (Barnes et al. 1973, Dudas and Pawluk 1976, Nelson et al. 1977, Matlick and Buseck 1978).

If 1:10 is used as the ratio of extrusive gas to solid weight (Mac Donald and Abbot 1970), then the 1868 eruption of Kilauea released $\sim10^{10}$ kg of gas. If 10% dry gas and a specific gravity of $\sim$1.25 kg/m^3 are assumed, calculations indicate that 1.25×10^9 m^3 would have been released. During the 1977 Kalalua eruption, a high mercury value of 200 μg/m^3 was measured downwind near the flank of the advancing lava front (Siegel and Siegel 1978a). If we assume (conservatively) that this sample was diluted 90% with clean air, the mercury content would be $\sim$2000 mg/m^3. Therefore, it can be estimated that 2.5 metric tons of mercury were released during this single eruptive episode at Kilauea. Application of a 20-fold factor for the 200-yr sum of Kilauea and Mauna Loa eruptions indicates that 500 metric tons of mercury were injected directly into the atmosphere during this period.

There is an additional contribution of mercury into the atmosphere: the continuous emission from cracks, fissures, and fumeroles. Outputs vary (Siegel et al. 1973, Siegel and Siegel 1978a, b, c), but these emissions can maintain localized high levels of mercury for centuries (Table 14.1). The total contribution from these unspectacular but virtually continuous emission sites is unknown, but aerial monitoring between the islands of Oahu and Hawaii up to an altitude of 2500 m has yielded mercury concentrations of 0.1–1.1 μg/m^3 (Siegel and Siegel 1978a). In a corridor 2500 m high, 1000 m wide, and 400,000 m long, the mercury content totaled $\sim$1 metric ton. Near fumeroles such as the Sulfur Bank, annual fallouts of mercury as high as 300 kg/km^2 have been reported (Siegel and Siegel 1978b). At a location 400 km to the northwest and normal to the tradewinds, an annual deposition of 30 kg/km^2 of mercury has been measured.

Although a complete model of the mercury budget of the Hawaiian Islands has not yet been constructed, many of the values for mercury are high compared to the generally accepted world or U.S. totals.

BIOACCUMULATION OF MERCURY

The object of the present study is to propose an important role for plants in the mercury cycle, especially with respect to atmospheric recycling. The emphasis thus far on nonbiogenic mercury sources has been necessary to determine a putative baseline of mercury. Before plants are considered as secondary atmospheric emission sources, they must first be examined as potential accumulators of mercury. Bioconcentration of mercury by plants has already been well documented (Siegel et al. 1975, 1980b). In contrast, other work shows that the mercury content in an area of Antarctica that is high in geothermal output is nevertheless depleted of surface mercury (McMurtry et al. 1979, Siegel et al. 1980b). The clayey mud and fine-grained volcanic sediments with and without algae in this area show a marked contrast in mercury content (Table 14.3).

For lichens and mosses, both generally known as accumulators of heavy metals (Garty et al. 1979, Rao et al. 1977, Richardson and Neiboer 1980), the plant:soil ratio for mercury concentration seems uniquely related to volcanic activity (Table 14.4) (after Siegel and Siegel 1976). Recent and active thermal zones show low plant:soil ratios for mercury, whether located in Hawaii or Iceland. In contrast, the ratios for lichens and mosses are consistently high in areas that have had only remote thermal histories. This difference can readily be explained when actual soil mercury levels are examined. On the average, recent and current thermal zones possess soil mercury levels 10 to 1000 times higher than nonthermal locations. Therefore, plants in nonthermal areas have retained their accumulative capacities, even after the input of mercury has diminished or ceased because of the reduction or cessation of geothermal activity.

Spatial proximity to a source of mercury is in itself a determinant of the mercury exposure potential, but a particular location—although known to be thermally active—may vary greatly in actual thermal events. Thus temporal proximity must also be considered. For example, in 1975 a station ~20 km east of Kilauea caldera along the East Rift (Island of Hawaii) was sampled for plant and soil mercury. The plants sampled included species of the genera *Dicranopteris, Nephrolepis, Cyperus, Metrosideros, Eucalyptus, Psidium,* and *Leucaena.* At the time of sampling, the aggregate mean plant mercury content was 353 (±63) μg/kg and the soil was 212 (±66) μg/kg, whereas plants in the Manoa district on the Island of Oahu, >350 km northwest of Kilauea, averaged 58 (±26) μg/kg and their soil 23 (±8) μg/kg. By 1981, the average for the same set of plant species at the Kilauea location had declined to 126 (±10) μg/kg and their soil to 63 (±2) μg/kg. The Manoa plant samples averaged 63 (±33) μg/kg.

What determined the marked decline in plant and soil mercury over the 6-yr interval? The 1975 analyses near Kilauea followed a period of extensive volcanism lasting from 1969 to late 1974, during which 12 eruptions generated $\sim$4.18 × 10^8 m^3 of magmatic extrusive material along with $\sim$4.5 × 10^9 m^3 of mercury-rich gases. The 1981 samplings followed a period of comparable length of time (1975–1981) but was marked by only three eruptions totaling $\sim$4.0 × 10^7 m^3 of solids and $\sim$4.0 × 10^8 m^3 of gas. Clearly, the mercury concentrations in plants and soils will vary with the amount of mercury generated by volcanic activity.

Among 275 vascular plant–soil (or substratum) sets studied for this paper, a new phenomenon was observed: a bimodal distribution of the concentration ratios. There were four widely separated areas sampled: Alaska (Juneau area) and "New England" (northern New York to southern Quebec), which are remote both in location and time from active volcanism, and Hawaii and Iceland, which are volcanically active locations (Table 14.5). The principal mode at all four locations was represented by an accumulation ratio of greater than two. At the "inactive" sites, however, a large fraction of plants (30–42%) have a smaller concentration of mercury in their tissues than the ambient substrate has. The same phenomenon exists among plants from "active" locations but is less marked. The most dramatic example of exclusion, that is, of plants lower in mercury than their soils, however, is from a highly mercuriferous site at the Puhimau thermal area on the Kilauea East Rift in Hawaii. *Portulaca sclerocarpa*, an endangered species, thrives in barren soil at root temperatures of 38–44°C. The plant mercury level is 100–300 μg/kg, whereas the level in the soil in which the plant is rooted is in the range of 3,000–17,000 μg/kg. Therefore, the plant:soil ratio is in the range of 0.006–0.07.

In marked contrast to the substantial body of information on bioaccumulation is the paradox of what is unmistakably "negative" accumulation or exclusion. There is no question about the importance of bioaccumulation as a means of moving mercury from soil to surface loci in concentrated form and from the soil into the food chain by herbivores. The case of "negative" accumulators, however, demands further consideration.

RELEASE OF VOLATILE MERCURY BY PLANTS

The original demonstration in 1975 of mercury concentration ratios that were appreciably <1.0 was taken as a strong indication of an exclusion mechanism (Siegel et al. 1975). This conclusion, although not unreasonable and still a possibility, is not supported by current experimental evidence. During the same period of time (1973–1975), it was found that the garlic vine *Pseudocalymma* released mercury at the remarkable rate of 39.5 µg/kg/hr at 25°C (Siegel et al. 1974). Two other species released more modest amounts of mercury, but the phenomenon was treated as a physiological oddity. It was not until some years later that the phenomenon was proven to be commonplace among both vascular and nonvascular plants (Kama and Siegel 1980, Siegel et al. 1980*a*). Among a broader selection of plants including vascular and nonvascular species, rates on the order of 0.5–1 µg/kg/hr (dry matter basis) were common (Table 14.6).

An area for further study may be to relate these observations to the "thermal" *Portulaca* plants exhibiting negative accumulation. *P. sclerocarpa* is an endemic species known to exist at only two or three sites in Hawaii, notably at Puhimau, a thermal area about 40 yr old on the Kilauea East Rift. At three sample stations these plants released substantial amounts of mercury at 25°C, although this species possesses typical reduced xerophytic leaves (Table 14.7).

Rock and soil surfaces also participate in mercury degassing (Desaeddeleer and Goldberg 1978, Siegel et al. 1980*a*), but there are clearly mechanistic distinctions between plant and soil processes. The mean thermal profile for mercury release from 10 soils is compared to the mean for 21 species of vascular and nonvascular plants and the highest releasing plant (garlic vine) in Figure 14.1. At temperatures at or above 25°C, the same limiting mechanism exists in plants and soils, and presumably this is the result of evaporation of mercury at these temperatures. From the data, a molar heat of activation of ∼14 kcal is obtained, corresponding to the heat of vaporization of mercury: 13.99 kcal/mol (Siegel et al. 1980*a*). Below 28°C, soil emission varies with temperature as it did above 28°C. In contrast, the slope for the plants increased substantially, indicating a greater temperature sensitivity. These results provide strong evidence that, although the release mechanism for barren soil is through simple evaporation, it is instead through an active or physiological process in plants.

Barber et al. (1973) concluded from their study of *Mentha* that higher plants neither accumulate nor release mercury. In our study, *Mentha* is also neither an accumulator nor a volatilizer of note. These authors also suggested that mercury in forest air arises from soil

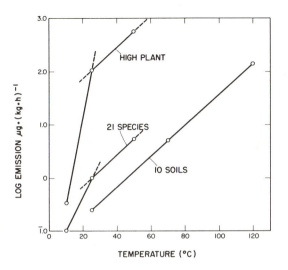

FIG. 14.1 Mercury emission as a function of temperature for plants and soils. "High plant" refers to the garlic vine, the most active mercury emitter found to date. The means for 21 plants include vascular and nonvascular forms, and the 10 soils include a number of sites in Hawaii and the continental United States.

microbial action. In addition to higher plants, algae, as exemplified by *Chlorella* (Ben-Bassat and Mayer 1975, 1977, 1978), can reduce and release mercury. Environmentally relevant release mechanisms have also been found in a study of plants growing on old volcanic and sedimentary surfaces in Hungary (Siegel et al. 1981).

Because of some original misinterpretations regarding the behavior of plants with respect to mercury, obvious explanations for plant—mercury phenomena must be viewed with caution. Thus transect data at the Kilauea Sulfur Bank fumerole (Table 14.8) would normally be treated as evidence for a mercury concentration gradient. A visual examination of the plant cover, however, shows no signs of a gradual increase in phytomass with distance and a concomitant decrease in mercury. The vegetative conditions are discrete units that can be defined as dense, barren, or isolated stands. An alternative hypothesis is that plants do not grow where substratum mercury is lower, but rather that the substratum mercury is lowered where the (colonizing) plants have grown.

By a judicious manipulation of numerical values from the wide-ranging capabilities of various plants (as discussed above), we have suggested elsewhere (Siegel and Siegel 1981) that in tropical volcanic zones mercury accumulated by plants can exceed 19 kg/km^2 and the annual mercury flux can exceed 530 kg/km^2. This is in contrast to the Environmental Protection Agency's calculation for the U.S. average annual degassing of 0.13 kg/km^2 (U.S. Environmental Protection Agency 1975). In the mixed climatic system of the continental United States, and with the assumption of an overall phytomass density of 10 kg/m^2 and a mean annual temperature of 10°C, the calculated mercury flux may be increased from 0.13 kg/km^2 to ~8.8 kg/km^2 annually.

Even this rate suggests that natural limits on the rate of degassing can be found in processes taking place in the soil and subsoil rather than in the vegetation per se, because these determine the supply of mercury available to the plant. It also remains possible that the Environmental Protection Agency's estimate of 0.13 kg/km^2 already includes the contribution of existing vegetation and would be far smaller if the continent were to be denuded of its plant cover. Plants as regulators of soil–air mercury distribution are obviously important. Their full significance remains to be established.

The metal-accumulating capabilities of plants have long found use in "geobotanical prospecting." Ordinarily, proximity to ore bodies in an area is reflected in the relative content of the metal of interest in the indicator species. In the case of volatile mercury, the indicator function may be applied not to the location of a mother lode but to the location and degree of volcanic or fumarolic degassing. This is shown dramatically in the distribution of mercury in *Equisetum arvense* plants around Mount St. Helens 1 yr after the eruption of May 18, 1980 (Siegel and Siegel 1982). The easterly direction of the plume from that massive eruption could be seen 1 yr later in the plants at Yakima and Toppenish to the east-northeast, which were 10–20 times richer in mercury than those at Portland to the southwest. Differences in soil mercury were less significant than that observed for shoots, suggesting an airborne but not ashborne source for the mercury.

ACKNOWLEDGMENTS

The authors wish to express their thanks to the following agencies for support during the past decade: National Aeronautics and Space Administration; National Science Foundation Division of Polar Programs; U.S. Department of Energy; NATO; National Institutes of Health, Division of Research Resources; the Cottrell Foundation;

American Lung Association; National Geographic Society; and the University of Hawaii Biomedical Research Support Program. Thanks also is given for advice, assistance, and support to Dr. G. Sigvaldasson, Nordic Institute of Volcanology, University of Reykjavik; Dr. R. Decker and the staff of USGS Hawaii Volcanoes Observatory; Hawaii Volcanoes National Park staff, National Park Service; and Dr. G. Eaton, USGS National Center, Reston, Virginia.

Above all, this paper is dedicated to the memory of the late Dr. Irving Bieger, USGS, whose inspiration started this study in 1970.

LITERATURE CITED

ANDREN, A., and J. NRIAGU. 1979. The global cycle of mercury. In *Biogeochemistry of Mercury in the Environment*, J. Nriagu (ed.), Elsevier/North-Holland Biomedical Press, Amsterdam/New York, pp. 1–21.

BARBER, J., W. BEAUFORD, and Y. SHIEH. 1973. Some aspects of mercury uptake by plant, algal, and bacterial systems in relation to biotransformation and volatilization. In *Mercury, Mercurials, and Mercaptans*, M. Miller, and T. Clarkson (eds.), Thomas, Springfield, Illinois, pp. 127–142.

BARNES, I., M. HINKLE, J. RAPP, C. HEROPOULOS, and W. VAUGHAN. 1973. Chemical composition of naturally occurring fluids in relation to mercury deposits in part of north-central California. *U.S. Geol. Surv. Bull. 138A*, 47 pp.

BEN-BASSAT, D., and A. MAYER. 1975. Volatilization of mercury by algae. *Plant Phys.* 38:128–132.

———. 1977. Reduction of mercuric chloride by Chlorella. *Physiol. Plant.* 40:157–162.

———. 1978. Light-induced mercury volatilization and oxygen evolution in Chlorella, and the effect of DCMO and methylamine. *Physiol. Plant.* 42:32–38.

CONNOR, J. 1979. Geochemistry of ohia and soil lichen, Puhimau thermal area Hawaii. *Sci. Total Environ.* 12:241–250.

DESAEDDELEER, G., and E. GOLDBERG. 1978. Rock volatility: Some initial experiments. *Geochem. J.* 12:75–79.

DUDAS, M., and S. PAWLUK. 1976. Nature of mercury in Chernozenic and luvisolic soils in Alberta. *Can. J. Soil Sci.* 56:413–423.

ESHLEMAN, A., S. SIEGEL, and B. SIEGEL. 1971. Is mercury from Hawaiian volcanoes a natural source of pollution? *Nature* 233:471–472.

GARTY, J., M. GALUN, and M. KESSEL. 1979. Localization of heavy metals and other elements in the lichen thallus. *New Phytol.* 82:159–168.

JOENSUU, O. 1971. Fossil fuels as a source of mercury pollution. *Science* 172:1027.

KAMA, W., and S. SIEGEL. 1980. Volatile mercury release from vascular plants. *Organic Geochem.* 2:99–101.

KRATKY, B., E. FUKUNAGA, J. HYLIN, and R. NAKANO. 1974. Volcanic air pollution: Deleterious effects on tomatoes. *J. Environ. Qual.* 3:17–22.

McCARTHY, J., L. MEUSCHKE, W. FICKLIN, and R. LEARNED. 1973. Mercury in the atmosphere. *U.S. Geol. Survey Prof. Paper* 713, pp. 37–41.

MacDONALD, G., and A. ABBOT. 1970. *Volcanoes from the Sea*. Univ. of Hawaii Press, Honolulu, pp. 56–75.

McMURTRY, G., R. BRILL, B. SIEGEL, and S. SIEGEL. 1979. Antarctic mercury distribution in comparison with Hawaii and Iceland. *Antarctic J. U.S.* 14:206–209.

MATLICK, J., and P. BUSECK. 1978. Exploration for geothermal areas using mercury: A geochemical technique. *Geotherm. Energy Magaz.* 6:18–23.

NELSON, H., B. LARSEN, E. JENNE, and D. LONG. 1977. Mercury dispersal from lode sources in the Kuskokwin River drainage, Alaska. *Science* 198:820−824.

RAO, D., G. ROBITAILLE, and F. LEBLANC. 1977. Influence of heavy metal pollution on lichen and bryophytes. *J. Hattori Biol. Lab.* 42:213−239.

RICHARDSON, D., and E. NIEBOER. 1980. Surface binding accumulation of metals in lichens. In *Cellular Interactions in Symbiosis and Parasitism*, C. Cook, P. Papa, and E. Rudolph (eds.). Ohio State Univ. Press, Columbus, pp. 75−94.

SIEGEL, B., and S. SIEGEL. 1973. Icelandic geothermal activity and the mercury of the Greenland icecap. *Nature* 241:526−527.

————. 1976. Unusual mercury accumulation in lichen flora of Montenegro. *Water Air Soil Polln.* 5:335−338.

————. 1978a. Mercury emission in Hawaii: An aerometric study of the Kalalua eruption of 1977. *Env. Sci. Technol.* 12:1036−1039.

————. 1978b. The Hawaii geothermal project: An aerometric study of mercury and sulfur emissions. *Geotherm. Res. Coun. Trans.* 2:596−599.

————. 1979. Biological indicators of atmospheric mercury. In *Biogeochemistry of Mercury in the Environment*, J. Nriagu (ed.). Elsevier/North-Holland Biomedical Press. pp. 131−149.

————. 1982. Mercury content of *Equisetum* plants around Mt. St. Helens one year after the major eruption. *Science* 216:292−293.

SIEGEL, S., and B. SIEGEL. 1975. Geothermal hazards: Mercury emission. *Env. Sci. Technol.* 9:473−474.

————. 1978c. Mercury fallout in Hawaii. *Water Air Soil Polln.* 9:113−118.

————. 1981. Vegetation and the recycling of mercury. Abstract in *The Role of Volcanic Emissions in Atmospheric Chemistry*, IAMP-CAGCP Session, Hamburg, Federal Republic of Germany, June 21−22.

SIEGEL, S., J. OKASAKO, P. KAALAKEA, and B. SIEGEL. 1980a. Release of volatile mercury from soils and non-vascular plants. *Organic Geochem.* 2:139−140.

SIEGEL, S., N. PUERNER, and T. SPEITEL. 1974. Release of volatile mercury from vascular plants. *Physiol. Plant.* 32:174−176.

SIEGEL, S., B. SIEGEL, A. ESHLEMAN, and K. BACHMAN. 1973. Geothermal sources and distribution of mercury in Hawaii. *Env. Biol. Med.* 2:81−89.

SIEGEL, S., B. SIEGEL, and G. McMURTRY. 1980b. Atmosphere−soil mercury distribution: The biotic factor. *Water Air Soil Polln.* 13:109−112.

SIEGEL, S., B. SIEGEL, and J. OKASAKO. 1981. A note on the anomalous mercury content of plants from the Lake Balaton region and its relation to release of mercury vapor. *Water Air Soil Polln.* 15:371−374.

SIEGEL, S., B. SIEGEL, N. PUERNER, T. SPEITEL, and F. THORARINSSON. 1975. Water and soil biotic relations in mercury distribution. *Water Air Soil Polln.* 4:9−18.

TUREKIAN, K., and F. WEDEPOHL. 1961. Distribution of the elements in some major units of the earth's crust. *Bull. Geol. Soc. Amer.* 72:175−192.

U.S. ENVIRONMENTAL PROTECTION AGENCY. 1975. *Material balance and technology assessment of mercury and its compounds on natural and regional bases.* Final Report, prepared for EPA Office of Toxic Substances by URS Research Co. EPA-560/3-75-007.

VOSSLER, T., D. ANDERSON, H. ARAS, J. PHELAN, and W. ZOLLER. 1981. Trace element composition of the Mount St. Helens plume: Stratospheric samples from the May 18 eruption. *Science* 211:827−830.

WOLLAST, R., B. BILLEN, and F. MACKENZIE. 1975. Behavior of mercury in natural systems and its global cycle. In *Ecological Toxicology Research: Effects of Heavy Metal and Organohalogen Compounds*, Proceedings of a NATO Science Conference, pp. 145−166.

TABLE 14.1
NATURAL GEOTHERMAL SOURCES OF MERCURY[a]

Source of mercury	Mercury content
Solids and liquids (μg/kg)	
Mean crustal	50
Basalt	
Soviet Union	6−1500
Iceland	67−815
Hawaii	20−1760
Antarctica (Erebus)	0−3.6
Tuff	
Soviet Union	trace−24,000
Iceland (Surtsey)	7−85
Hawaii	350−375
Tephra	
Antarctica (Erebus)	0−3.6
Iceland (Surtsey)	350
Sublimates	
Hawaii	16−1700
Antarctica	1200−6250
Condensates	
Soviet Union	0.2−72

	Range	Mean	s.e.
Gases (μg/m³)			
Fumerolic			
Iceland	1.3−37.0	10.0	5.2
Hawaii	1.0−40.7	17.6	6.1
Volcanic	-		
Iceland	4.8−9.6	7.1	0.7
Hawaii	0.7−200	29.2	7.9
Antarctica (Erebus)	3.6−24.1	11.3	6.4
Nonthermal areas			
Iceland	0.62−1.0	0.8	0.2
Hawaii	0.04−1.3	1.1	0.5
California	0.01	0.01	
New York	0.014	0.014	
East Pacific	0.001	0.001	

[a]Data from McCarthy et al. 1973; McMurtry et al. 1979; Siegel et al. 1973, 1981; and Turekian and Wedepohl 1961.

TABLE 14.2
SURFACE:ATMOSPHERE MERCURY DISTRIBUTION RATIOS
IN THREE VOLCANIC REGIONS

Soil:Air mercury		
Site	Ratio	Number of sets
Iceland	10.8	36
Hawaii	6.8	220
Antarctica	0.7	32

TABLE 14.3
EVIDENCE FOR BIOACCUMULATION OF MERCURY IN ANTARCTIC SAMPLES

Sample	Mercury (μg/kg)
Muds (clayey sediments)	
Without algae	2.0−7.1
Algae rich	10.8−13.0
Igneous fines (basalt and ash)	
Without algae	1.2−3.1
Algae rich	9.0−46.2

TABLE 14.4
Plant:Soil Concentration Ratios for Mercury in Relationship to Current Volcanism

Sample site	Plant:soil mercury
Recent volcanic surfaces	
Usnea (Kilauea East Rift, Hawaii)	3.0
Cladonia (Mount Hekla, Iceland)	0.05
Dicranella (Kilauea East Rift, Hawaii)	1.0
Rhacomitrium (Kilauea East Rift, Hawaii)	4.0
	(Average 2.0)
Active fumerolic areas	
Cladonia (Puhimau Crater Area, Hawaii)	1.8
Haematoma (Sulfur Bank Area, Hawaii)	6.8
Dicranella (Kilauea East Rift, Hawaii)	1.2
Polytrichum (Kilauea East Rift, Hawaii)	1.1
	(Average 2.9)
Inactive and nonvolcanic areas	
Usnea (Conifer forest, Alaska)	27.0
Cladonia (Mount Greylock, Massachusetts)	30.0
Haematoma (Dinaric Alps, Montenegro)	90.5
Dicranella (Glacial morraine, southeast Iceland)	21.2
Polytrichum (Glacial morraine, Alaska)	42.8
Rhacomitrium (Glacial morraine, southeast Iceland)	7.9
	(Average 36.6)

TABLE 14.5
Bioconcentration Modalities in Vascular Plant:Soil Ratios[a]

Plant:soil mercury ratio	Percentage of samples of that ratio			
	Iceland (79)[b]	Hawaii (22)	New England (38)	Alaska (66)
< 0.1	0	12	32	30
0.1−0.67	10	18	10	0
0.67−1.33	0	13	0	5
1.33−2	14	3	18	10
> 2	76	54	40	55

[a]After Siegel et al. 1975.

[b]Numbers in parentheses indicate sample size.

TABLE 14.6

VOLATILE RELEASE RATES FOR SELECTED PLANTS

Vascular species	Rate (μg/kg/hr)[a]	Nonvascular species	Rate (μg/kg/hr)[a]
Aloe	0.85 ± 0.05	Lichens	
Pineapple	0.82 ± 0.03	*Usnea*	0.66
Coconut	0.24	*Cladonia*	0.12 ± 0.03
Cypress	0.34	*Stereocaulon*	0.84
Staghorn fern	0.73 ± 0.27		
Hibiscus	0.49	Mosses	
Leucaena	3.05 ± 0.30	*Leucobryum*	2.05
Pandanus	1.59 ± 0.39	*Bryum*	0.40 ± 0.05
Avocado	4.13 ± 0.22	*Polytrichum*	0.40
Mangrove	0.10 ± 0.04	Fungi	
Sugarcane	0.65 ± 0.04	*Polyporus*	1.20
Papyrus	0.70	*Fomes*	2.43 ± 1.20
		Trametes	0.93 ± 0.48

[a]Dry matter basis.

TABLE 14.7

MERCURY DATA FOR *Portulaca sclerocarpa* GRAY OF
PUHIMAU, HAWAII, AUGUST 30, 1980

Site	Mercury content (g/kg)		Mercury release (ng/kg/hr)
	Plant	Soil	
NE 4	313	4286	735
0, 0	99	16610	294
SW 4	97	3173	445
(Averages)	(169)	(8023)	(491)

TABLE 14.8
PLANT:SOIL MERCURY RELATIONSHIPS ALONG A TRANSECT AT
KILAUEA SULFUR BANK FUMEROLE, HAWAII

Distance from steam vent (m)[a]	Condition	Mercury in substratum (μg/kg)
+75	Dense vegetation	32
+60	Isolated stands	9
+50	Isolated stands	16
+30	Isolated stands	13
+10	Barren red clay and sulfur	405
0	Barren gypsum and silica	450
−10	Barren red clay	290
−15	Dense vegetation	56

[a] + denotes positions southwest and − denotes positions northeast of the steam vents along the transect line.

Appendix

Organization and Conduct of Ecological Research Programs in the Vicinity of Mount St. Helens

Jerry F. Franklin

ABSTRACT

Ecological scientists interested in conducting research at Mount St. Helens after the eruption faced logistical and legal hurdles that prevented easy access to the volcano. Also, coordination among scientists was necessary to ensure effective field studies and to protect research sites. A variety of successful responses dealt with these problems, including support of early reconnaissance and planning efforts by the Pacific Northwest Forest and Range Experiment Station, research field sampling sessions, encouragement of coordinating efforts by the National Science Foundation, and creation of the St. Helens Forest Land Research Cooperative. Ecological research at Mount St. Helens has been active and productive; however, funding has not been adequate to allow full advantage of research opportunities. Improved communication between geologists and ecologists, coordination of research projects (including integrated ecosystem investigations), and protection of important research sites are considerations for future research on Mount St. Helens.

INTRODUCTION

The eruptions of Mount St. Helens provided unique opportunities for ecological research in a relatively accessible region and created a huge field laboratory in which the following question could be addressed: What happens when 100,000 ha of forests and meadows are covered with varying depths and textures of volcanic tephra and when all the above-ground plant and animal life on adjacent clearcut and virgin forest lands is destroyed? Basic theories of primary and secondary succession, population biology, ecosystem recovery, and land−water interactions could be tested on a previously unimaginable scale. Also, ecologists could provide essential technical information to land managers and planners interested in natural recovery rates, artificial revegetation, and erosion control.

Research on and around Mount St. Helens created a number of special challenges. Many scientists were interested in initiating studies that required funding and access to the affected areas. Organizations potentially responsible for conducting or funding research quickly expressed concern about the coordination and duplication of research efforts. Land managers were faced with promptly screening large numbers of scientific requests for access to the volcano and ensuring that not only the needs of basic research but their own requirements for information were addressed during the development of scientific programs.

Logistics and restricted access to areas near the volcano created other concerns. A special access permit was necessary for the restricted "red zone" (the most endangered area subject to closure during threatened or actual volcanic activity), and entry required special equipment (e.g., radios and four-wheel drive vehicles), had restricted time limits, and was influenced by the constant possibility of rapid eviction whenever further eruptions threatened. Since expensive helicopters were the only means by which to reach much of the devastated region, adequate funds, weather conditions, and limited air space around the volcano had to be considered. The large geographical extent of the affected area made it difficult to select a manageable number of sites for scientific studies. Finally, recognition that many important but short-lived phenomena were occurring required prompt initiation of scientific observations.

Responding to these challenges, the scientific community, institutions, and organizations developed procedures for organizing and coordinating the ecological research program at Mount St. Helens which may provide useful guides for future situations. Several important factors contributed to the success of these procedures: (1) research programs (as well as their administration) were flexible, because altered physical conditions, restricted access, legal limitations, funding levels, and experience required a more evolutionary approach; (2) the scientific community received extensive cooperation from the Gifford Pinchot National Forest and the Weyerhaeuser Company; (3) cooperation among scientists contributed to research successes; and (4) a basic administrative infrastructure for interdisciplinary research—National Science Foundation sponsored ecosystem programs—provided a basis for a coordinated approach to research at Mount St. Helens.

INITIAL CHALLENGES

Challenges to conducting research on Mount St. Helens generally fell into three categories: (1) logistical and legal access to areas in the vicinity of the volcano, (2) coordination and duplication of research efforts, and (3) cooperative efforts.

LOGISTICAL AND LEGAL ACCESS

During 1980 and 1981, access to study sites in the red zone around Mount St. Helens was legally restricted and limited to helicopters. As time went on, the area of the red zone decreased in size, rules governing research effort in the zone eased, and ground access (some of which was possible almost immediately following the eruption) increased greatly. Nonetheless, restrictions on access during the first 18 months following the eruption created significant constraints on research objectives, although institutions controlling access considered them necessary to protect the safety of involved scientists.

Access to the red zone was controlled by the Gifford Pinchot National Forest, the Washington State Department of Emergency Services, and the local county sheriff departments. The U.S. Geologic Survey provided these agencies with information concerning volcanic hazards. During the summer of 1980, blanket entry permits for multiple trips were generally not issued, and scientists with valid research projects had to be certified before they were issued entry permits by the Emergency Coordination Center of the Gifford Pinchot National Forest. Also, during that summer, the red zone was closed when volcano warnings were issued or when direct observation of the mountain by airplanes or remote television cameras was restricted due to inclement weather.

Activity inside the red zone was governed by regulations, most of which were made to ensure the safety of researchers. During the first two summers following the eruption, regulations required that researchers stay within 15 min walking time from a helicopter or ground vehicle in the red zone and 30 min in the blue zone. No dropoffs or overnight stays were permitted in the red zone. Furthermore, all parties were required to maintain radio contact with the Volcano Control Center in Vancouver, Washington, or with a patrolling aircraft. Alternatively, scientists could be in radio contact with a base station outside the restricted zone that, in turn, was in phone contact with the Volcano Control Center. Obtaining necessary USDA Forest Service radios was often difficult, because purchase of radios with appropriate crystals was restricted and only a limited number of radios were available for loan. These rules were periodically modified as perception of the volcano and its hazards changed.

Access to the red zone required helicopters during the summer of 1980. These were frequently unavailable (especially on short notice), expensive when available, and restricted by poor weather. Permission to enter the red zone by helicopter could be denied because of excessive traffic in the restricted airspace. By1981, however, reconstructed roads permitted access to most of the northwestern and eastern portions of the blast zone, and much of the remainder of the zone had roads by the fall of 1982. Helicopters might be necessary to provide transportation for some research projects for many years to come on Mount St. Helens, on Mount Margaret and associated peaks, and in parts of the upper North Fork Toutle River basin.

COORDINATION AND DUPLICATION OF RESEARCH EFFORTS

Since there was considerable duplication of interests among scientists, agencies funding scientific studies and land management organizations responsible for the accommodation of scientists wanted to limit research duplication. They hoped that efforts by various scientists on similar research topics would be coordinated, and thus everyone's needs for information would be met. Plans and information were initially exchanged, thereby preventing accidental duplications in proposals and providing opportunities to develop compatible or joint studies.

Independent site selection by each scientist threatened to clutter the landscape with plots, creating duplication and reducing the certainty of adequate documentation. It became essential to identify a series of common sampling sites in the blast zone and in the extensive region of tephra deposition to the northeast to provide the best possible investigative efficiency, site preservation, and ease of access. Data gathered at these sites could be shared among projects, thereby satisfying common needs and integrating the studies. Direct comparisons of results were possible. In addition, land management agencies could more easily protect carefully chosen sites.

One major tactic supporting coordination and logistical efficiency in research was the intensive field sampling sessions or "pulses" during several 2-week periods in 1980 and 1981. Research in the restricted zones on Mount St. Helens was facilitated by provision of living and working accommodations, equipment, ground vehicles, and helicopters and by an easing of restrictions on permits. These sessions were organized by the Pacific Northwest Forest and Range Experiment

Station (PNW Station) personnel, were financed by the National Science Foundation (NSF) and the PNW Station, and were based at the Cispus Environmental Center located about 16 km south of Randle, Washington, and 32 km northeast of Mount St. Helens. The first pulse was from September 7 to 20, 1980, and involved about 125 scientists; participation averaged 40 persons per day. The second pulse was from August 23 to September 5, 1981, involved about 170 scientific personnel, and averaged 60 persons per day.

The pulses had four major objectives. First, regulations on access into the red zone could be dealt with more efficiently. It was advantageous to both researchers and administrators to have one person act as pulse coordinator to handle clearances, obtain radios, and check procedures for as many as 12 research teams. Second, a scarce resource—helicopters—could be used more efficiently. With many scientists working simultaneously at Mount St. Helens, helicopter missions could be contracted for continuous availability at reduced rates, which allowed for more efficient scheduling. The ability of many scientists to schedule their sample collection and plot establishment during this common period saved time and money and reduced logistical concerns. It also made it possible for administrators to accommodate more scientists in the area than would otherwise have been possible.

The third and fourth objectives of the pulses were coordination and information sharing, respectively. Coordination between two or more research teams was facilitated by simultaneous field sampling sessions on the volcano. Information sharing took many forms, including regular evening "show-and-tell" sessions during which each team reported on hypotheses, accomplishments, and difficulties. Many research projects were altered and alternative hypotheses generated during these discussions.

COOPERATIVE EFFORTS

Scientists and scientific organizations joined to minimize logistical problems and coordinate research programs. Officials of the Gifford Pinchot National Forest and the Pacific Northwest Region 6, parts of the National Forest Systems branch of the USDA Forest Service, recognized a need for a group of ecologists to screen biological research proposals. The PNW Station, a unit of the research branch of the USDA Forest Service, became the coordinating and certifying body for biological and management research in the restricted zones. An in-house group of subject matter coordinators was established to assist in evaluating proposals. Once a request for research within the restricted zone was certified by the PNW Station, necessary approvals were provided by Pacific Northwest Region 6 and permits were issued by the Gifford Pinchot National Forest.

In late May of 1980, the NSF and PNW Station organized a meeting of approximately 30 interested ecological scientists from several universities to discuss possible research projects on Mount St. Helens. At this meeting, scientists became acquainted, shared plans for research projects, and pooled information on access, conditions within the area, and funding. Dr. David Johnson, representing NSF, announced that the Foundation would expedite the consideration of short proposals for new grants or supplements to conduct research on transient phenomena at Mount St. Helens. As promised, paperwork was minimal and these emergency grants were awarded promptly.

When access to the blast areas became possible in July, the PNW Station sponsored a series of reconnaissance flights in large helicopters for 8 to 10 scientists. Following these flights, meetings to plan research programs and develop common sampling strategies were held for groups of researchers interested in the effect of tephra deposition on ecosystems, terrestrial ecology of the blast zone, lake ecology, and stream ecology and erosion.

The St. Helens Forest Land Research Cooperative was established for communication and cooperation among the three major landowners in the affected area: the Washington State Department of Natural Resources, the USDA Forest Service, and the Weyerhaeuser Company. Land managers and scientists were included in their deliberations. A major activity of the Cooperative was the "Technical Needs Workshop" held on September 4–5, 1980, and November 4, 1981, in Olympia, Washington. The objective of the workshop was to identify technical information required for the rehabilitation of volcanically impacted lands around Mount St. Helens. Attendees included scientists from major universities in Oregon and Washington and from sponsoring organizations. Although the Cooperative had only limited ability to ensure that its recommendations were carried through, it continued to encourage information exchange among scientists and between the scientific community and forest land managers to promote implementation of rehabilitation.

INSTITUTIONAL RESPONSES

Institutional responses to opportunities for scientific research at Mount St. Helens were generally rapid. Some delay did occur, however, while administrators of the restricted zones developed procedures for a totally novel situation.

The Division of Environmental Biology of the NSF responded to the unusual opportunity by providing modest amounts of money for critically important early studies (critical since the landscape, with many transient phenomena, was changing rapidly). Twenty grants or supplements were made by August 7, 1980. The NSF also furthered research by encouraging collaboration and information exchange and by providing a grant for helicopter support to the scientific community.

PNW Station developed a research plan for Mount St. Helens during the first week after the May 18 eruption. A supplemental appropriation by the U.S. Congress provided approximately $600,000, of which about one third went for grants and contracts to non–Forest Service scientists, one third for in-house research, and one third for logistical support.

The Weyerhaeuser Company has a substantial staff at the Forestry Research Center in Centralia, Washington, and thus was able to promptly direct efforts toward timber salvage and reforestation of its lands. The earliest reforestation and grass seeding studies were initiated by company personnel who also began erosion, watershed, wildlife, and other research projects. Weyerhaeuser continued to cooperate with Mount St. Helens researchers and periodically updated its master list of research projects.

SUBSEQUENT PROCEDURES

Thirty months after the eruption, research at Mount St. Helens began to follow more normal procedures as ground access to the blast zone became greatly improved because of the advent of timber salvage sales and activities by the U.S. Army Corps of Engineers. Helicopters continued to be essential for research in the areas of Mount St. Helens, Spirit Lake, and Mount Margaret. Restricted zones continued to shrink. There was also a reduction in the administrative complexity associated with scientific permits and regulations for the red zone. Improved access was a mixed blessing, because it permitted activities that could have had significant negative impacts on research opportunities and increased problems associated with vandalism and general public use.

Coordination among scientific groups continued to be important. A vegetation-oriented group met to share information and plans for research. An interdisciplinary research group, funded by NSF and led by L. C. Bliss (Chairman of the Botany Department at the University of Washington) focused on effects of the eruptions on timberline vegetation. At an ad hoc workshop, chaired by J. Winjum of the Weyerhaeuser Company and Bliss, scientists met in October, 1981, to discuss the need for additional research and coordination and concluded that a central repository for publications and other information was needed. Subsequently, the Washington State Library at Olympia was selected as the repository site. Communication between the geological and ecological researchers was inadequate despite the critical role that geological events and substrate played and continue to play ecologically. A workshop was held in May, 1982, under the joint sponsorship of the USDA Forest Service and NSF to facilitate information exchange between the two groups of scientists. The St. Helens Forest Lands Research Cooperative continued a series of technical information workshops in 1982 for land managers.

Funding from all sources for Mount St. Helens research became much more limited than it was during the first 6 months following the eruption. General budget reductions and the lack of supplemental appropriations for the volcanic research were major factors. Neither PNW Station nor the NSF could continue to provide general helicopter support or to organize and fund research pulses, and Mount St. Helens research began to compete with other programs and topics for continued support.

When the amount of the investment is considered, Mount St. Helens ecological research has been very productive to date, although only time will determine its value to basic science. Much information has been generated, with some surprising findings concerning revegetation and linkages between organisms and their environment. These findings have begun to appear in the scientific literature and in this volume.

Scientific data and results were quickly utilized by land managers. A number of informal communication channels developed to meet the managers' needs for the best current information to develop plans for protection, timber salvage, and revegetation of the volcanically affected region. In the USDA Forest Service, area ecologists, staff specialists at the Gifford Pinchot National Forest, and silviculturalists on the Ranger Districts all functioned as intermediaries in a two-way flow of information. PNW Station personnel improved normal (and typically slow) reporting processes by providing frequent briefings on the latest findings for National Forest personnel. The forestry research staffs of the Weyerhaeuser Company and the Washington State

Department of Natural Resources continued to work closely with their respective foresters near Mount St. Helens. Throughout, the objective has been to ensure that the best available technical knowledge provided, and will continue to provide, the basis for management decisions.

CONTRIBUTORS

A. B. Adams
Department of Botany
University of Washington
Seattle, WA 98195

Joseph A. Antos
Department of Botany and Plant
    Pathology
Oregon State University
Corvallis, OR 97331

D. E. Bilderback
Department of Botany
University of Montana
Missoula, MT 59812

L. C. Bliss
Department of Botany
University of Washington
Seattle, WA 98195

Jerry J. Bromenshenk
Department of Zoology
University of Montana
Missoula, MT 59812

Virginia H. Dale
Department of Forest Science
Oregon State University
Corvallis, OR 97331-5704

Roger del Moral
Department of Botany
University of Washington
Seattle, WA 98195

William H. Emmingham
Willamette National Forest
Eugene, Oregon 97401

D. G. Fellin
Forest Science Laboratory
USDA Forest Service
Missoula, MT 59806

Jerry F. Franklin
Pacific Northwest Forest and Range
    Experiment Station
USDA Forest Service
Corvallis, OR 97331

R. R. Gilchrist
SW Washington Region
Weyerhaeuser Company
Longview, WA 98632

Miles A. Hemstrom
Department of Forest Science
Oregon State University
Corvallis, OR 97331

Arthur R. Kruckeberg
Department of Botany
University of Washington
Seattle, WA 98195

D. A. Leslie
Western Forestry Research Center
Weyerhaeuser Company
Centralia, WA 98531

Richard N. Mack
Department of Botany
Washington State University
Pullman, WA 99164

Arthur McKee
Department of Forest Science
Oregon State University
Corvallis, OR 97331

Joseph E. Means
Pacific Northwest Forest and Range
    Experiment Station
USDA Forest Service
Corvallis, OR 97331

William H. Moir
Southwestern Region
USDA Forest Service
Albuquerque, NM 87102

Elizabeth M. W. Pincha
1100 S.W. Eastbrook Road
Normandy Park, WA 98166

R. C. Postle
Division of Mining and Reclamation
  Enforcement with Hopi Tribe
Kykotsmovia, AZ 86039

H. E. Reinhardt
College of Arts and Sciences
University of Montana
Missoula, MT 59812

B. Z. Siegel
Pacific Biomedical Research Center
University of Hawaii at Manoa
Honolulu, HI 96822

S. M. Siegel
Department of Botany
University of Hawaii at Manoa
Honolulu, HI 96822

J. H. Slone
Department of Botany
University of Montana
Missoula, MT 59812

R. G. Stevens
Western Forestry Research Center
Weyerhaeuser Company
Centralia, WA 98531

Frederick Swanson
USDA Forest Service
Forest Sciences Laboratory
3200 Jefferson Way
Corvallis, OR 97331

Stephen D. Veirs, Jr.
National Park Service
Redwood National Park
Arcata, CA 95521

J. K. Winjum
Western Forestry Research Center
Weyerhaeuser Company
Centralia, WA 98531

G. M. Yamasaki
USDA Forest Service
Ogden, UT 84401

Donald B. Zobel
Department of Botany and Plant
  Pathology
Oregon State University
Corvallis, OR 97331

INDEX

Abies spp., 104; *A. amabilis*, 9, 11, 76, 78, 86, 88, 91, 101, 102, 103, 105, 170, 174, 177, 191, 192, 193, 231, 235, 284; *A. grandis*, 276, 284; *A. lasiocarpa*, 9, 12, 284; *A. procera*, 12, 76, 78, 86, 88, 95, 103, 170, 191, 213

Acer circinatum, 191, 193, 233; *A. macrophyllum*, 233, 235, 236, 239

Acheta domesticus, 303

Achillea millefolium, 13; var. *lanulosa*, 157

Achlys triphylla, 233

Aerial reconnaissance, 24, 25, 28, 34, 42, 43, 44, 149, 170, 172, 174, 189, 341

Agropyron smithii, 248; *A. spicatum*, 263, 268, 269, 270, 273

Agrostis spp., 87; *A. diegoensis*, 13, 152, 154, 155, 157; *A. interrupta*, 273

Alaska cedar, 12

Alder, 9, 12, 34, 61, 315

Alectoria sarmentosa, 87

Algae, 266, 320; on arid steppes, 273, 274, 275; blooms, 275; blue-green, 5; green, 273; release of mercury by, 327; yellow-green, 273

Alluvium: plant burial in, 246, 247, 249; soil composition of, 253

Alnus rubra, 75, 76, 78, 81, 86, 88, 101, 103, 104, 105; *A. sinuata*, 12, 91, 95, 101, 102, 105, 231, 233

Alpine collomia, 15

Alpine habitats, 6, 8, 12, 13, 15

Alpine heather, 13

Alpine saxifrage, 15

Anaphalis margaritacea, 105, 174, 233

Andropogon gerardi, 248

Anemone occidentalis, 8

Antennaria spp., 159; *A. microphylla*, 13

Ants, 304

Apis millifera, 303

Arctostaphylos nevadensis, 10, 95, 154; *A. uva-ursi*, 10

Arenaria pumicola, 8

Argyroxiphium, 6

Arnica viscosa, 8

Artemisia spp., 263, 266; *A. rigida*, 269, 270; *A. tridentata*, 263, 268, 269, 270, 272, 273, 274

Artificial revegetation, 210, 213, 214, 219, 223

Ash. *See* Ashfall(s)

Ashfall(s): algal growth on, 273, 274, 275; alteration of soil seed bank, 271, 272; application to foliage, 307; in artificial diets, 308, 309; characteristics, 38, 78, 307; damage to vegetation, 54, 189; deposition on conifer needles, 283, 284, 285, 288, 289; in eastern Washington, 304; effect of previous depositions of, 263, 264; effect on bees, 304, 305; effect on community stability, 265, 266; effect on epiphytic fungi, 282, 287; effect on flowering and fruiting, 268, 270, 271; effect on insects, 304, 305; effect on leaves, 269; effect on needle carbohydrates, 282; effect on parasites, 310, 315; effect on photosynthesis, 282; effect on plant heating, 237, 266; effect on pollen assemblages, 263, 264; effect on pollination, 268; effect on pollinators, 304, 305; effect on respiration, 282; effect on seedlings, 273; effect on shrubs,

233, 234, 235, 236, 237; survival patterns, 236, 237, 238; on temperate volcanoes, 6; on tropical volcanoes, 6; vegetative changes, 188; vegetative cover, 213, 218; on volcanoes in general, 3, 5; western spruce budworm on, 302. *See also* Floras; Plant succession; Trees; Vegetation types

Plantago patagonica, 273

Plant cover, 194, 195, 198, 199, 203, 204, 210, 214, 215

Plant life form, as survival influence, 188, 194

Plant recovery. *See* Plant succession

Plant seedlings. *See* Plants; Seedling(s)

Plant succession: on all disturbance habitats, 105; in blast zone, 86, 87, 91, 191; in blowndown areas, 174, 175; in Cascades, 70; in clearcut areas, 174, 175; climax, 11; effects of erosion on, 102; effects of snow avalanches on, 102; factors affecting, 70, 72, 188, 189, 191, 193, 194, 195, 198, 199, 204, 205; after glacial recession, 70, 71, 87, 93, 98, 101; on Mount St. Helens, 93; on mudflows, 87, 93, 98, 101; in 1981, 87; in 1984, 87; patterns of, 71, 72, 73, 95; processes of, 72; secondary, 222; and soil dynamics, 70; species recruitment for, 95, 97; successional models of, 103, 104; after volcanic eruptions, 70, 71, 87, 88, 91, 101. *See also* Plants

Poa spp., 273; *P. pratensis*, 265; *P. sandbergii*, 269, 270

Pogonomyrmex owyheei, 272

Polemonium elegans, 8

Polygonum majus, 273; *P. newberryi*, 8, 15, 152, 154, 155, 158

Ponderosa pine, 282, 284, 285, 287, 288, 289, 295, 296

Populus trichocarpa, 12, 78, 87, 88, 104, 105

Portulaca sclerocarpa, 325, 326

Potentilla, 159; *P. gracilis*, 269, 270, 273

Prunus virginiana, 268

Pseudocalymma, 320, 326

Pseudotsuga mengiesii, 9, 76, 78, 81, 86, 88, 102, 103, 104, 105, 170, 191, 193, 213, 231, 234, 282, 284

Psidium, 324

Psylla pyricola, 304

Pumice, 149, 158, 161, 250; communities, 9; flora on, 12, 151, 155; habitat, 13, 15, 16; soil composition from, 4; substrate, 12; thermal characteristics of, 254. *See also* Tephra

Pussy Paws, 15

Puu Paui Volcano, 223

Pyroclastic deposits habitat, 13, 46

Pyroclastic flows, 1, 29, 43, 48, 71, 72, 95, 97, 98, 99, 100, 149, 150, 169; germination trials on, 159; impact on vegetation, 191; physical structure of, 100; plant cover on, 168, 170, 172, 173; plant diversity on, 97; plant recovery on, 178; plant succession on, 95, 97, 170, 191; soil analysis of, 73; vegetation damage by, 151, 160

Quartz Creek, 47, 48

Red pine, 298

Redwood, 249

Red Zone, 36, 40, 337, 338, 339, 340, 342

Reithrodontomys megalotis, 272

Research Natural Areas, 11

Rhacomitrium spp., 101; *R. canescens*, 100; *R. lanuginosum*, 100

Rhododendron albiflorum, 192

Riparian areas: after eruption, 170; plant cover, 176, 178; prior to eruption, 170; recolonization, 177; species composition, 177; species richness, 168, 178; vegetation,

Designer: U.C. Press Staff
Compositor: Trend Western
Text: 10/12 Baskerville
Display: Baskerville
Printer: Braun-Brumfield, Inc.
Binder: Braun-Brumfield, Inc.